普通高等教育“十四五”规划教材

金属材料专业英语

Specialty English for Metal Materials

李玉超　张冬梅　查俊伟　主编

扫码查看本书
数字资源

扫码查看本书
单词、词组

北　京
冶　金　工　业　出　版　社
2023

内 容 提 要

本书为大学本科专业英语阅读教材，是按照五位一体思路编撰的与金属材料专业密切相关的专业英语教材，兼顾知识的系统性和实用性。全书共包含六个部分（Part），主要内容包括金属材料介绍、金属合金、机械加工、热处理、金属材料的性能、金属材料的应用。每个部分（Part）的知识体系又拓展为4～5个单元（Unit），共计28个单元（Unit）。

本书可作为高等院校金属材料等相关专业英语教学用书，也可供从事金属材料、机械加工等科研与设计人员阅读参考。

图书在版编目(CIP)数据

金属材料专业英语/李玉超，张冬梅，查俊伟主编．—北京：冶金工业出版社，2023.1

普通高等教育“十四五”规划教材

ISBN 978-7-5024-9368-4

Ⅰ.①金…　Ⅱ.①李…　②张…　③查…　Ⅲ.①金属材料—英语—高等学校—教材　Ⅳ.①TG14

中国国家版本馆 CIP 数据核字(2023)第 022818 号

金属材料专业英语

出版发行　冶金工业出版社　　**电　　话**　(010)64027926
地　　址　北京市东城区嵩祝院北巷39号　　**邮　　编**　100009
网　　址　www. mip1953. com　　**电子信箱**　service@ mip1953. com

责任编辑　刘林烨　美术编辑　彭子赫　版式设计　郑小利
责任校对　郑　娟　责任印制　禹　蕊
三河市双峰印刷装订有限公司印刷
2023年1月第1版，2023年1月第1次印刷
787mm×1092mm　1/16；16印张；387千字；242页
定价 49.00 元

投稿电话　(010)64027932　投稿信箱　tougao@cnmip. com. cn
营销中心电话　(010)64044283
冶金工业出版社天猫旗舰店　yjgycbs. tmall. com
(本书如有印装质量问题，本社营销中心负责退换)

前　言

本书是在完成教育部规定的大学英语课程基础上，根据不同专业所开设的一门专业英语提高课程。专业英语提高课以阅读为主，该课程的设置对本科生培养至关重要，它承上启下：一方面它是大学英语的进一步延伸，偏重于加强专业词汇、阅读和撰写科技文献的能力；另一方面，它又是不同专业课程学习的补充，通过英语文献的阅读拓展学生国际化视野，有利于培养综合性人才。

金属材料在人类社会中的使用历史悠久，领域广泛。金属材料既是制备民用产品的基础材料，也是制造飞机、高铁、卫星、航母等高端装备的关键材料。各高校金属材料工程专业既需要培养金属材料工程理论和专业知识扎实、实践能力强，熟悉金属材料选材、加工、结构表征、腐蚀与防护、性能检测等的研究型人才，又需要培养能够在材料、冶金、机械等领域从事金属相关工艺研究、技术开发、生产管理、质量检测及一线工作的应用型人才。在国际化、新工科及OBE教育理念的大背景下，作者针对高等院校金属材料类相关专业英语教材相对匮乏、专业性不强的情况下编写了本书。本书具有以下特点。

（1）本书按照金属结构、合金类型、金属加工工艺、金属热处理、金属性能及应用五位一体的思路编撰出的与金属材料专业知识密切相关，适合金属材料专业本科人才培养的专业英语教材。

（2）本书以学生为中心，以教学需求为前提，由浅入深，循序渐进，重点培养学生的认知和阅读能力，促进金属材料专业本科生综合素质和国际化水平的提高。

（3）每个单元包含一篇800字左右的正文和一篇1200字左右的阅读材料。由简入深，阅读材料是对正文内容的进一步补充，相得益彰。

（4）本书兼顾知识的系统性和实用性，大部分章节知识配有相关示意图，以满足学习需要。

（5）总结了各章节重要专业词汇和短语，几乎涵盖了整个金属材料知识体系的重要内容，以便于读者自学。

(6) 本书有配套视频、电子课件及思维导图等资源，以供学习者参考。

本书共分6个部分（Part），28个单元（Unit）。全书由聊城大学李玉超、张冬梅协同统稿、翻译、讲解，由北京科技大学查俊伟教授主审。本书编写分工为：第一部分（Part 1）由聊城大学李玉超、张冬梅编写，第二部分（Part 2）由聊城大学赵性川、王长征、李玉超编写，第三部分（Part 3）由聊城大学李伟、郝雪卉、张冬梅编写，第四部分（Part 4）由山东科技大学张芬、宋亮编写，第五部分（Part 5）由北京科技大学查俊伟、聊城大学李玉超、张召编写，第六部分（Part 6）由聊城大学张冬梅、西北工业大学苗应刚编写。

本书在编写过程中，参考了相关国外著作和文献资料，并获得了聊城大学规划教材建设项目（编号JC201903）的基金资助。本书由聊城大学材料科学与工程学院李玉超教授团队与北京科技大学查俊伟教授团队相关成员联合编写，同时得到了聊城大学滕谋勇教授和西北工业大学苗应刚教授的大力支持。在此，向各方为本书出版做出贡献的人士表示衷心感谢！

由于编者水平所限，书中不妥之处，敬请广大读者批评指正。

编　者
2022年4月3日

目　　录

第一部分　金属材料介绍
Part 1　Introduction to Metallic Materials

第二部分 金属合金
Part 2 Metal Alloys

第三部分　机械加工
Part 3　Mechanical Machining

第四部分 热处理
Part 4 Heat Treatment

第五部分　金属材料的性能
Part 5　Properties of Metallic Materials

第一部分　金属材料介绍

Part 1　Introduction to Metallic Materials

单元1　金属介绍
Unit 1　Introduction to Metals

扫码查看
讲课视频

1.1　教学内容
1.1　Teaching Materials

A metal(from Greek μεταλλον métallon,"mine,quarry,metal") is a material that,when freshly prepared,polished, or fractured, shows a lustrous appearance, and conducts electricity and heat relatively well. Metals are typically malleable (they can be hammered into thin sheets) or ductile(can be drawn into wires). A metal may be a chemical element such as iron; an alloy such as stainless steel; or a molecular compound such as polymeric sulfur nitride.

In physics,a metal is generally regarded as any substance capable of conducting electricity at a temperature of absolute zero. Many elements and compounds that are not normally classified as metals become metallic under high pressures. For example,the nonmetal iodine gradually becomes a metal at a pressure of between 40 and 170 thousand times atmospheric pressure. Equally,some materials regarded as metals can become nonmetals. Sodium,for example,becomes a nonmetal at pressure of just under two million times atmospheric pressure.

In chemistry,two elements that would otherwise qualify(in physics) as brittle metals—arsenic and antimony—are commonly instead recognised as metalloids,on account of their predominately non-metallic chemistry. Around 95 of the 118 elements in the periodic table are metals. The number is inexact as the boundaries between metals,nonmetals,and metalloids fluctuate slightly due to a lack of universally accepted definitions of the categories involved.

In astrophysics,the term "metal" is cast more widely to refer to all chemical elements in a star that are heavier than the lightest two,hydrogen and helium,and not just traditional metals. A star fuses lighter atoms,mostly hydrogen and helium,into heavier atoms over its lifetime. Used in that sense,the metallicity of an astronomical object is the proportion of its matter made up of the heavier chemical elements.

Metals,as chemical elements,comprise 25% of the Earth's crust and are present in many aspects of modern life. The strength and resilience of some metals has led to their frequent use in,for example, high-rise building and bridge construction, as well as most vehicles, many home appliances,tools,pipes,and railroad tracks. Precious metals were historically used as coinage,but in the modern era,coinage metals have extended to at least 23 of the chemical elements.

The history of refined metals is thought to begin with the use of copper about 11000 years ago. Gold,silver,iron,lead,and brass were likewise in use before the first known appearance of bronze in the 5th millennium BCE. Subsequent developments include the production of early forms of steel; the discovery of sodium—the first light metal—in 1809; the rise of modern alloy steels; and, since the end of World War Ⅱ, the development of more sophisticated alloys.

Metals are mostly solid, crystalline, hard, strong, and dense. They are shiny and lustrous, at least when freshly prepared, polished, or fractured. Metals are typically malleable and ductile, deforming under stress without cleaving. The nondirectional nature of metallic bonding is thought to contribute significantly to the ductility of most metallic solids. The electronic structure of metals means they are relatively good conductors of electricity. Electrons in matter can only have fixed rather than variable energy levels, and in a metal the energy levels of the electrons in its electron cloud, at least to some degree, correspond to the energy levels at which electrical conduction can occur. Metals are relatively good conductors of heat. The electrons in a metal's electron cloud are highly mobile and easily able to pass on heat-induced vibrational energy. The contribution of a metal's electrons to its heat capacity and thermal conductivity, and the electrical conductivity of the metal itself can be calculated from the free electron model.

Metals are usually inclined to form cations through electron loss. Most will react with oxygen in the air to form oxides over various timescales (potassium burns in seconds while iron rusts over years). Some others, like palladium, platinum and gold, do not react with the atmosphere at all. The oxides of metals are generally basic, as opposed to those of nonmetals, which are acidic or neutral. Painting, anodizing or plating metals are good ways to prevent their corrosion. However, a more reactive metal in the electrochemical series must be chosen for coating, especially when chipping of the coating is expected.

Metals can be categorised according to their physical or chemical properties. Categories below include ferrous and non-ferrous metals; brittle metals and refractory metals; white metals; heavy and light metals; and base, noble, and precious metals. The Metallic elements table categorises the elemental metals on the basis of their chemical properties into alkali and alkaline earth metals; transition and post-transition metals; and lanthanides and actinides. Other categories are possible, depending on the criteria for inclusion. For example, the ferromagnetic metals—those metals that are magnetic at room temperature—are iron, cobalt, and nickel.

(Source: *Metal*, Wikipedia)

Words and Expressions

quarry *n.* 采石场，沙石场
polish *n.* 上光剂；*vt.* 抛光
lustrous *a.* 有光泽的
malleable *a.* 可锻造的，有延展性的
ductile *a.* 可延展的
iodine *n.* [化] 碘酒，碘 (I)
sulfur *n.* [化] 硫，硫黄；*v.* 用硫黄处理
nitride *n.* 氮化物
arsenic *n.* [化] 砷 (As)，砒霜
antimony *n.* [化] 锑 (Sb)
metalloid *n.* 类金属
metallicity *n.* 金属物，金属性
resilience *n.* 恢复力，弹力
appliance *n.* 家用电器，装置
coinage *n.* 造币，铸币
hydrogen *n.* [化工氢 (H)]，氢气

helium *n.* ［化］氦（He）
copper *n.*［化］铜（Cu）
gold *n.*［化］金（Au）
silver *n.*［化］银（Ag）
bronze *n.* 青铜
brass *n.* 黄铜
iron *n.*［化］铁（Fe）
lead *n.*［化］铅（Pb）
sodium *n.*［化］钠（Na）
crystalline *a.* 结晶的，似水晶的
electron *n.* 电子
covalently *a.* 共价的
silicon *n.*［化］硅（Si）
corrosion *n.* 腐蚀，腐蚀产生的物质
cation *n.* ［化］正离子
palladium *n.*［化］钯（Pd）
platinum *n.*［化］铂（Pt）
anodizing *n.* 阳极氧化
plating *n.* 电镀
compound *n.* 化合物
coating *n.* 涂层，包衣，衣料
ferromagnetic *a.* 铁磁的，铁磁体的
refractory *n.* 耐火材料
vibrational *a.* ［力］振动的，震动性的
chemical element 化学元素
stainless steel 不锈钢
precious metals 贵金属
light metal 轻金属
alloy steel 合金钢
energy level 能级
electron cloud 电子云
heat capacity 热容，热容量
thermal conductivity 导热系数，热导率
energy gap 能隙
ferrous metal 黑色金属，铁系金属
nonferrous metal 有色金属

Phrases

regard as 认为……，被视为……
on account of 由于，为了……的缘故
lack of 没有，缺乏，不够
refer to 参考，涉及
lead to 导致
correspond to 相当于……，符合于……
incline to 想要，有……倾向

Notes

(1) In physics, a metal is generally regarded as any substance capable of conducting electricity at a temperature of absolute zero.

regarded as 译为“被认为是……”；capable of 译为“能够……”。

参考译文：物理学上，金属通常被认为是任何能够在绝对零度下导电的物质。

(2) In chemistry, two elements that would otherwise qualify (in physics) as brittle metals—arsenic and antimony—are commonly instead recognised as metalloids, on account of their predominately non-metallic chemistry.

recognised as 译为“被认为是……”；on account of 译为“由于……，为了……的缘故”。

参考译文：化学上，砷和锑这两种（在物理学中）原本属于脆性金属的元素通常被认为是类金属，主要因为它们是非金属化学结合的。

(3) Metals are mostly solid, crystalline, hard, strong, and dense. They are shiny and lustrous,

at least when freshly prepared, polished, or fractured.

freshly 译为“最新的，最初的”；at least 译为“至少”。

参考译文：金属大多是结晶的、坚硬、坚固、致密的固体。它们是有光泽的，至少在最初制备、抛光或断裂时是如此。

(4) Electrons in matter can only have fixed rather than variable energy levels, and in a metal the energy levels of the electrons in its electron cloud, at least to some degree, correspond to the energy levels at which electrical conduction can occur.

rather than 译为“而不是……”；to some degree 译为“在一定程度上”。

参考译文：物质中的电子只能有固定的而不是可变的能级，在金属内部，电子云中电子的能级至少在一定程度上对应于能够发生电导的能级。

(5) Painting, anodizing or plating metals are good ways to prevent their corrosion. However, a more reactive metal in the electrochemical series must be chosen for coating, especially when chipping of the coating is expected.

good ways to 译为“好的方法”；expected 译为“期望的，预期的，预料的”。

参考译文：涂漆、阳极氧化或镀层是防止金属腐蚀的好方法。因此，对于电化学系列中较活跃的金属就必须选择涂层，特别是在预料会出现涂层碎屑时。

1.2 阅读材料

1.2 Reading Metarials

History and Background of Metallic Materials

1. Early Origins

Modern humans as Homo sapiens have inhabited the globe for at least 300000 years, but a mere blink in geological time. The early acquaintance with metals, no doubt in the form of native gold, detected as tiny glittering nuggets in river streams, may well date back tens of thousands of years. Similarly, native copper occurring in surface outcroppings of mineralisation will also have attracted human attention. Eventually people learnt that collected pieces of such metals could be easily hammered and fashioned, in the case of gold, into pleasing shapes and ornaments. But gold being so soft would have had no practical use for early people. Native copper, on the other hand, hardens under such hammering and working and gradually practical uses for this new material were also found. The chalcolithic (copper stone) period of human development gradually evolved.

There was no significant growth in the practical use of metals until, slowly over thousands of years, humans learnt to control fire and to generate the high temperatures necessary to smelt copper from those same exposed areas of mineralisation which until then had yielded only native copper metal. Recent archaeological evidence has emerged, indicating that early developments in the use of native metals, and eventually smelting, seem to have occurred in a variety of locations around the world and did not necessarily spread outwards from one single source. One of the most remarkable discoveries of early smelting was in 1965 at Timna in present-day Israel, just north of Eilat and west of the Wadi Arabah, which marks the border with Jordan. Excavations revealed evidence of primitive copper smelting operations originally dated to around 4000 B. C. More recent research indicates the site may have been used for smelting as early as the 7th millennium BC. Other early evidence of copper mining and smelting some 6500 years ago has been uncovered at Rudna Glava in the Balkans. This is probably one of the sources of copper used by the Vinca people of that region, who made a wide variety of copper utensils, weapons and ornaments.

It was the advent of the Bronze Age sometime between 4000 and 5000 years ago that rapidly promoted the manufacture and use of copper alloy artefacts and weaponry. Some of the early "accidental bronzes" did not actually contain tin but antimony or arsenic no doubt derived from mixed orebodies. The true Bronze Age is, however, firmly associated with tin. Quite where the first tin came from is still a matter of conjecture. Tin is not a widely occurring metal. Iberia, Italy, Saxony and rather later Cornwall may have been early sources of this metal. It is also quite possible that trade may have developed in tin from as far afield as Nigeria, South-East Asia and China.

2. Early Metal Trading

The early history of serious metal trading almost certainly started with the Bronze Age. Demand for bronze burgeoned as soon as its remarkable properties had been recognised and developed. Today we can only speculate as to how such trade developed.

Early overland movements of goods of all descriptions, in the Near East for example, would usually have been by pack horses, mules, donkeys and camel caravans. Such caravans and similar movements would have needed armed protection and frequently would have paid tribute or taxes at different stages on their way to their ultimate destination. For countries with access to the Black Sea or bordering the Mediterranean, small coasters would have been the preferred method of transporting goods. It is very likely that such coastal trade, rather like the inland caravans, would have resulted in goods changing ownership at many different trading centres or ports en route. Over centuries, patterns of trade will have so developed that the ultimate recipient of the metals may have had no idea as to its origins. Nor would the smelter and caster of the ingots have known the ultimate destination. Periodically, there were no doubt interruptions to supply, often through local wars or natural disasters. So trade was inevitably haphazard, unreliable and certainly risky.

Early trade would have been conducted frequently by barter but there is ample evidence of various units of gold, silver and even lead for low values being used as a form of currency, along with precious stones, which in turn could also be used for profitable barter. Dependency on copper and bronze for a wide range of utensils, tools, weaponry and ornaments began gradually, from about 1500 B. C. onwards, to give way to the use of iron. But the rise of the use of iron, as smithing and tempering skills improved, was increasingly in the forefront of new developments, especially in weaponry and tools. Usage of both copper and lead continued to flourish under the Romans, especially for roofing and pipes and various utensils. The more expensive bronze gradually fell out of favour but continued to be used for bell-making and other decorative rather than practical purposes.

3. The not Quite so Dark Ages

Compared with the use of iron and early forms of steel, the principal non-ferrous metals tended to languish. Growth slowed. Leaping through the centuries, the collapse first of the incredibly long-lived Egyptian empire and the later retreat and fall of the Roman Empire led to the so-called Dark Ages. More than a thousand years were to elapse after the withdrawal of the Romans from Northern and Western Europe, including Britain, before non-ferrous metals were to regain some of the importance, they had enjoyed during the Bronze Age.

The centuries after the retreat of the Romans were not quite so dark as is often supposed and certainly not destitute of developments in mining. Among the Saxons there were mining specialists well versed in smelting and metal production and already exploiting deposits in their home and neighbouring territories. They brought their techniques to England during this period and redeveloped the tin and copper mines of Cornwall and Devon, and the lead mines of Derbyshire and along the Pennines. The old trading routes were gradually reopened, with tin and lead the principal products. But, until the beginning of the nineteenth century, the production of base metals across Europe was on a small and localised scale.

4. The Industrial Revolution

In 1709, Abraham Darby had already successfully experimented with coal's derivative coke for the smelting of iron and had brought down the costs dramatically. James Watt some 50 years later famously set about improving Newcomen's remarkable engine and in 1769 he and Matthew Boulton patented the world's first true steam engine. There followed a succession of ingenious and practical industrial developments. By the end of the eighteenth century the Industrial Revolution was well underway with Britain leading the world. Steam power rapidly transformed all aspects of British "manufacturing" and Britain's resources in copper, lead and to a lesser extent tin were also now rapidly developed to feed the demand for engineering products needed for the new "Machine Age". The second half of the nineteenth century was a period of innovation, industrial development and change on an unprecedented scale. The development of electric power and lighting was initiated by the ingenious experiments of the visionary Michael Faraday. Samuel Morse's first telegraph line in 1844 led to a rapid expansion in demand for copper wire for this purpose, soon also for the transmission of electric power. Before long Alexander Graham Bell's telephone and Thomas Edison's electric light were beginning to penetrate into factories, offices and homes on a massive scale.

Throughout most of the eighteenth and nineteenth centuries Britain, as earlier mentioned, had been at the forefront of innovation and industrial production. First the canals and then the railways were soon followed by the first iron-clad steamships. These transformed the methods of transportation of both goods and people. Along with these developments Britain's consumption of copper, lead, tin and zinc had so dramatically expanded that its own mine resources were soon outstripped by demand. Already by the early nineteenth century Britain was importing increasing quantities of copper ores and concentrates from Chile, Cuba and Spain to feed the principal smelters in Swansea. UK copper production in the first decade of the century had been two-thirds of the entire world production, which was then estimated at a mere 12000 tons. By 1900 world mine production had rocketed to 494000 tons but that of Britain had declined to less than 1000 tons. From being an exporter of surplus metal early in the nineteenth century, Britain had become a massive importer.

From the perspective of the interested observer at the beginning of the twenty-first century, the most astonishing aspect of the metal business a century earlier was the huge importance of lead and secondly zinc, which both far outweighed in tonnage terms the usage of copper. World production of lead at the beginning of the nineteenth century was already more than double that of copper, and this was still broadly the case even by the beginning of the twentieth century. Lead's low melting temperature, easy castability and malleable properties encouraged its use, especially in Britain, for roofing, guttering, water pipes and sheet and pipe for the fast-growing chemical industry. It was also virtually the only material suitable for sheathing power cables in the early days of electric power transmission.

(Source: Martin Thompson. *Base Metals Handbook*. Cambridge: Woodhead Publishing Limited, 2006)

Words and Expressions

mineralisation *n.* 矿化作用
chalcolithic *a.* 铜石并用时代的，红铜时代的
smelting *n.* ［冶］熔炼，冶炼
excavation *n.* （对古物的）发掘，挖掘
penetrate *v.* 刺入，穿透
outstrip *v.* 超过，胜过
concentrate *n.* ［矿］精矿，浓缩液
pipe *n.* 管道，管子
roofing *n.* 屋顶，盖屋顶
zinc *n.* ［化］锌（Zn）
sheet *n.* 板材
sheathing *n.* 防护物，外壳
far afield 远离；广泛地
en route （法）在途中

1.3 问题与讨论

1.3 Questions and Discussions

(1) What basic properties do metals have?
(2) How to distinguish metals from nonmetals?
(3) Why metals are usually good conductors?
(4) List some categories of metals according to their physical and chemical properties.
(5) The good ways to prevent corrosion.

单元 2　金属种类
Unit 2　Types of Metals

扫码查看
讲课视频

2.1　教学内容
2.1　Teaching Materials

1. Classification of Metals

A large number of metals are available in nature. More than eighty different types of metals find use today. They can be classified in a variety of ways depending on what property or characteristic you use as a yardstick.

(1) Classification by Iron Content

The most common way of classifying them is by their iron content. When a metal contains iron, it is known as a ferrous metal. The iron imparts magnetic properties to the material and also makes them prone to corrosion. Almost 90 percent of manufactured metals are ferrous metals. Metals that do not have any iron content are non-ferrous metals. These metals do not possess any magnetic properties. Examples include but are not limited to aluminum, lead, brass, copper and zinc.

(2) Classification by Atomic Structure

They may also be classified based on their atomic structure according to the periodic table. When done, a metal may be known as alkaline, alkaline earth, or a transition metal. Metals belonging to the same group behave similarly when reacting with other elements. Thus, they have similar chemical properties.

(3) Magnetic and Non-Magnetic Metals

Another way to differentiate metals is by looking how they interact with magnets. It is possible to divide metals as magnetic and non-magnetic on that basis. While ferromagnetic metals attract strongly to magnets, paramagnetic ones only show weak interactions. Lastly, there is a group called diamagnetic metals that rather show a weak repulsion to magnets.

2. Iron, Its Alloys and Their Properties

(1) Iron

It would not be an exaggeration to refer to iron as the lifeblood of our civilisation. Approximately 5 percent of the Earth's crust is iron. Thus, it is an incredibly easy metal to find. Pure iron is an unstable element though. At the first opportunity, it reacts with the oxygen in the air to form iron oxide. Extracting iron from its ores uses a blast furnace. Pig iron is achieved from the first stage of the blast furnace which can be further refined to obtain pure iron.

(2) Steel

Pure iron is stronger than other metals, but it leaves much to be desired. For one, pure iron is not resistant to corrosion. Secondly, it is also extremely heavy due to its high density. Adding carbon to iron alleviates these weaknesses to a certain extent. This mixture of iron and carbon up to specified limits is known as carbon steel. Over 3500 grades of steel are available today. It has high tensile strength and a high strength-to-weight ratio. This allows usage of steel parts and components that are small in size but still strong. Steel is also extremely durable. This means a steel structure can last longer and withstand external factors better than other alternatives. It is also ductile and can be shaped into required forms without compromising its properties. Carbon steels often contain other elements to improve certain properties. Like chromium for corrosion resistance or manganese to improve hardenability and tensile strength.

Alloy steels contains multiple elements to enhance various properties. Metals such as manganese, titanium, copper, nickel, silicon, and aluminium may be added in different proportions. This improves steel 's hardenability, weldability, corrosion resistance, ductility and formability. Applications for alloy steels are electric motors, bearings, heating elements, springs, gears, and pipelines.

3. Non-ferrous Metals

In addition to ferrous metals, we have a large selection of non-ferrous ones. Aluminium, copper, titanium, zinc and lead are commonly seen. Each has certain qualities that make them useful in different industries.

(1) Aluminum

Aluminium derives primarily from its ore bauxite. It is light, strong and functional. It is the most widespread metal on Earth and its use has permeated applications everywhere. This is because of its properties such as durability, light weight, corrosion resistance, electrical conductivity and ability to form alloys with most metals. It also doesn't magnetise and is easy to machine.

(2) Copper

Copper and its alloys have a long history because it is easy to form. Even today, it is an important metal in the industry. It does not occur in nature in its pure form. Metals are good conductors and copper stands out more than the others. Due to its excellent electrical conductivity, it finds application in electrical circuits as a conductor. Its conductivity is second only to silver. It has also excellent heat conductivity. This is why many cooking utensils are from copper. Brass is an alloy of copper and zinc. Brass is a great candidate for low friction applications such as locks, bearings, plumbing, musical instruments, tools and fittings. Bronze is also an alloy of copper. But instead of zinc, bronze contains tin. Bronze is brittle, hard, and resists fatigue well. It also has good electrical and thermal conductivity and corrosion resistance.

(3) Titanium

Titanium is an important engineering metal due to being strong and lightweight. It also has high thermal stability even at temperatures as high as 480 degrees Celsius. Due to these properties, it finds application in the aerospace industry and military equipments. Since titanium is also corrosion resistant, titanium is also used in medical applications and chemical and sporting goods industry.

(Source: Andreas Velling. *Types of Metal—Pure Metals. Alloys&Their Applications*. Fractory Website)

Words and Expressions

yardstick　*n.* 尺度，准绳
aluminum　*n.*［化］铝（Al）
alloy　*n.* 合金
lifeblood　*n.* 生机的根源，命脉；
alleviate　*v.* 减轻，缓和
durable　*a.* 耐用的
chromium　*n.*［化］铬（Cr）
manganese　*n.*［化］锰（Mn）
hardenability　*n.* 淬透性，可硬化性
weldability　*n.* 焊接性，可焊性
formability　*n.* 成形性，可锻性
ductility　*n.* 延展性，柔韧性
titanium　*n.*［化］钛（Ti）
vanadium　*n.*［化］钒（V）
permeate　*v.* 渗透，弥漫
bauxite　*n.* 矾土，铝土矿
magnetise　*n.* 磁化
circuit　*n.*［电］电路
friction　*n.* 摩擦，摩擦力
plumbing　*n.* 管路系统，管道设备
fitting　*n.* 配件，附件
tin　*n.*［化］锡（Sn）
fatigue　*n.* 疲劳
lightweight　*n.* 轻量级，轻质
spring　*n.* 弹簧
pipeline　*n.* 输送管道，渠道
weakness　*n.* 虚弱，缺点，缺陷
Celsius　*n.* 摄氏度
periodic table　周期表
alkali metal　碱金属
alkaline earth metal　碱土金属
transition metal　过渡金属
iron oxide　氧化铁
diamagnetic metals　抗磁性金属
pig iron　生铁
corrosion resistance　抗腐蚀性
carbon steel　碳钢
tensile strength　拉伸强度
electrical conductivity　电导率
thermal stability　热稳定性

Phrases

not limited to　不限于
in addition to　除……外
depend on　取决于……，依赖，依靠
impart to　赋予，传授，给予
based on　基于，以……为基础
belong to　属于，归属
resistant to　对……有抵抗力的，耐……
due to　由于，应归于
shaped into　成型
stand out　突出，站出来
second to　次于，仅次于……的
instead of　代替，而不是

Notes

(1) When a metal contains iron, it is known as a ferrous metal. The iron imparts magnetic properties to the material and also makes them prone to corrosion.

known as 译为“被熟知的，被认为是……”；impart to 译为“赋予……性质”。

参考译文：当一种金属含有铁时，它被称为黑色金属。铁赋予这种材料磁性，也使其容易腐蚀。

(2) It is possible to divide metals as magnetic and non-magnetic on that basis. While ferromagnetic metals attract strongly to magnets, paramagnetic ones only show weak interactions.

It is possible to…译为“可以……，做某事是可能的”；attract to 译为“吸引”。

参考译文：可以将金属分为磁性金属和非磁性金属。铁磁金属对磁铁的有很强的吸引力，而顺磁性金属的相互作用很弱。

(3) For one, pure iron is not resistant to corrosion. Secondly, it is also extremely heavy due to its high density. Adding carbon to iron alleviates these weaknesses to a certain extent.

resistant to 译为“抵抗……，耐……”；due to 译为“由于……原因”；to a certain extent 译为“在一定程度上”。

参考译文：首先，纯金属不耐腐蚀；其次，由于高密度导致非常重。在铁中加入碳在一定程度上缓解了这些弱点。

(4) Carbon steels often contain other elements to improve certain properties. Like chromium for corrosion resistance or manganese to improve hardenability and tensile strength.

certain 表示“一些，某个”，也可以表示“特定的”。

参考译文：碳素钢还含有其他元素以改善某些特定性能。如铬用于提高耐蚀性或锰提高淬透性和抗拉强度等。

(5) Metals are good conductors and copper stands out more than the others. Due to its excellent electrical conductivity, it finds application in electrical circuits as a conductor. Its conductivity is second only to silver.

stands out 这里指“突出的，出众的”；second to 表示“仅次于”。

参考译文：金属是良导体，铜最为突出。其优良的导电性使它作为导体在电路中得到应用，它的导电性仅次于银。

2.2　阅读材料

2.2　Reading Materials

Metals, Metalloid, and Nonmetals

The elements can be classified as metals, nonmetals, or metalloids. Metals are good conductors of heat and electricity, and are malleable (they can be hammered into sheets) and ductile (they can be drawn into wire). Most of the metals are solids at room temperature, with a characteristic silvery shine (except for mercury, which is a liquid). Nonmetals are (usually) poor conductors of heat and electricity, and are not malleable or ductile; many of the elemental nonmetals are gases at room temperature, while others are liquids and others are solids. The metalloids are intermediate in their properties. In their physical properties, they are more like the nonmetals, but under certain circumstances, several of them can be made to conduct electricity. These semiconductors are extremely important in computers and other electronic devices.

On many periodic tables, a jagged black line (Figure 1) along the right side of the table separates the metals from the nonmetals. The metals are to the left of the line (except for hydrogen, which is a nonmetal), the nonmetals are to the right of the line, and the elements immediately adjacent to the line are the metalloids.

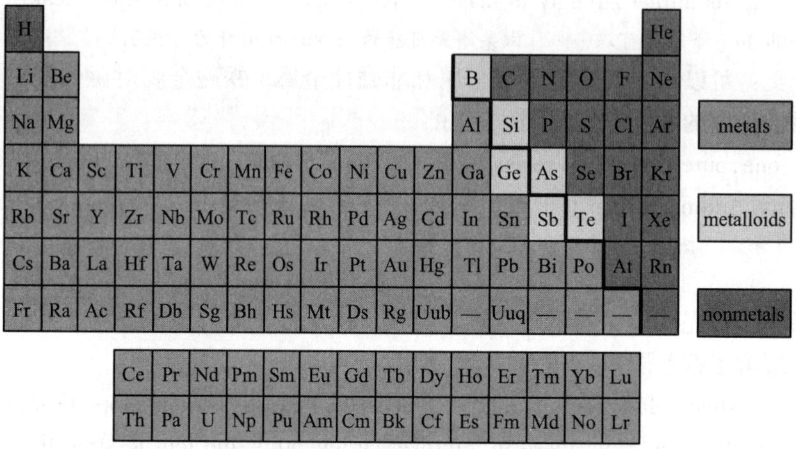

H																	He
Li	Be											B	C	N	O	F	Ne
Na	Mg											Al	Si	P	S	Cl	Ar
K	Ca	Sc	Ti	V	Cr	Mn	Fe	Co	Ni	Cu	Zn	Ga	Ge	As	Se	Br	Kr
Rb	Sr	Y	Zr	Nb	Mo	Tc	Ru	Rh	Pd	Ag	Cd	In	Sn	Sb	Te	I	Xe
Cs	Ba	La	Hf	Ta	W	Re	Os	Ir	Pt	Au	Hg	Tl	Pb	Bi	Po	At	Rn
Fr	Ra	Ac	Rf	Db	Sg	Bh	Hs	Mt	Ds	Rg	Uub	—	Uuq	—	—	—	—

Ce	Pr	Nd	Pm	Sm	Eu	Gd	Tb	Dy	Ho	Er	Tm	Yb	Lu
Th	Pa	U	Np	Pu	Am	Cm	Bk	Cf	Es	Fm	Md	No	Lr

metals
metalloids
nonmetals

Figure 1 Periodic table

When elements combine to form compounds, there are two major types of bonding that can result. Ionic bonds form when there is a transfer of electrons from one species to another, producing charged ions which attract each other very strongly by electrostatic interactions, and covalent bonds, which result when atoms share electrons to produce neutral molecules. In general, metal and nonmetals combine to form ionic compounds, while nonmetals combine with other nonmetals to form covalent compounds (molecules).

Since the metals are further to the left on the periodic table, they have low ionization energies and low electron affinities, so they lose electrons relatively easily and gain them with difficulty. They also have relatively few valence electrons, and can form ions (and thereby satisfy the octet rule) more easily by losing their valence electrons to form positively charged cations.

(1) The main-group metals usually form charges that are the same as their group number: that is, the Group 1A metals such as sodium and potassium form 1+ charges, the Group 2A metals such as magnesium and calcium form 2+ charges, and the Group 3A metals such as aluminum form 3+ charges.

(2) The metals which follow the transition metals (towards the bottom of Groups 4A and 5A) can lose either their outermost s and p electrons, forming charges that are identical to their group number, or they can lose just the p electrons while retaining their two s electrons, forming charges that are the group number minus two. In other words, tin and lead in Group 4A can form either 4+

or 2+ charges, while bismuth in Group 5A can form either a 5+ or a 3+ charge.

(3) The transition metals usually are capable of forming 2+ charges by losing their valence s electrons, but can also lose electrons from their d orbitals to form other charges. Most of the transition metals can form more than one possible charge in ionic compounds.

Nonmetals are further to the right on the periodic table, and have high ionization energies and high electron affinities, so they gain electrons relatively easily, and lose them with difficulty. They also have a larger number of valence electrons, and are already close to having a complete octet of eight electrons. The nonmetals gain electrons until they have the same number of electrons as the nearest noble gas (Group 8A), forming negatively charged anions which have charges that are the group number minus eight. That is, the Group 7A nonmetals form 1-charges, the Group 6A nonmetals form 2-charges, and the Group 5A metals form 3-charges. The Group 8A elements already have eight electrons in their valence shells, and have little tendency to either gain or lose electrons, and do not readily form ionic or molecular compounds.

Ionic compounds are held together in a regular array called a crystal lattice by the attractive forces between the oppositely charged cations and anions. These attractive forces are very strong, and most ionic compounds therefore have very high melting points. For instance, sodium chloride, NaCl, melts at 801℃, while aluminum oxide, Al_2O_3, melts at 2054℃. Ionic compounds are typically hard, rigid, and brittle. Ionic compounds do not conduct electricity, because the ions are not free to move in the solid phase, but ionic compounds can conduct electricity when they are dissolved in water.

When nonmetals combine with other nonmetals, they tend to share electrons in covalent bonds instead of forming ions, resulting in the formation of neutral molecules. (Keep in mind that since hydrogen is also a nonmetal, the combination of hydrogen with another nonmetal will also produce a covalent bond.) Molecular compounds can be gases, liquids, or low melting point solids, and comprise a wide variety of substances.

When metals combine with each other, the bonding is usually described as metallic bonding. In this model, each metal atom donates one or more of its valence electrons to make an electron sea that surrounds all of the atoms, holding the substance together by the attraction between the metal cations and the negatively charged electrons. Since the electrons in the electron sea can move freely, metals conduct electricity very easily, unlike molecules, where the electrons are more localized. Metal atoms can move past each other more easily than those in ionic compounds (which are held in fixed positions by the attractions between cations and anions), allowing the metal to be hammered into sheets or drawn into wire. Different metals can be combined very easily to make alloys, which can have much different physical properties from their constituent metals. Steel is an alloy of iron and carbon, which is much harder than iron itself; chromium, vanadium, nickel, and other metals are also often added to iron to make steels of various types. Brass is an alloy of copper and zinc which is used in plumbing fixtures, electrical parts, and musical instruments. Bronze is an alloy of copper and tin, which is much harder than copper; when bronze was discovered by ancient civilizations, it marked a significant step forward from the use of less durable stone tools.

(Source: kevin A. *Boudreaux*, *Metals*, *Metalloids*, *and Nonmetals*. AUS Website)

Words and Expressions

tendency　*n.* 偏好，趋势，倾向
semiconductor　*n.* 半导体
affinity　*n.* 喜好，匹配度，亲和力
potassium　*n.* [化] 钾（K）
attraction　*n.* 吸引力，引力
jag　*n.* 缺口；*vt.* 使成锯齿状，使成缺口
chloride　*n.* 氯化物
melting point　熔点
valence electron　价电子

2.3　问题与讨论
2.3　Questions and Discussions

(1) What is the most common way of classifying metals?
(2) How to distingwish metals by atomic structure?
(3) What is ferrous metals?
(4) What is the difference of copper, brass, and bronze?
(5) Introduce some of nonferrous metals.
(6) What is alloy steel and its applications?

单元3 冶 金 学
Unit 3 Metallurgy

扫码查看
讲课视频

3.1 教学内容
3.1 Teaching Materials

Modern civilization started with the discovery of metals. Without metals it would not have been possible to build railway, transport, bridges, buildings, cars, generate power or any of the automobile, electrical and electronic industries. As the standard of living grows and the sheer size of population growth coupled with ongoing requirements in the developed world has created unprecedented demand for metals. More metals have been extracted in the twentieth century than the entire amount produced from the beginning of mankind's history until 1900 A. D.

Metallurgy is a domain of materials science and engineering that studies the physical and chemical behavior of metallic elements, their intermetallic compounds, and their mixtures, which are called alloys. Metallurgy encompasses both the science and the technology of metals. That is, the way in which science is applied to the production of metals, and the engineering of metal components used in products for both consumers and manufacturers. Metallurgy is distinct from the craft of metalworking. Metalworking relies on metallurgy in a similar manner to how medicine relies on medical science for technical advancement.

The science of metallurgy is subdivided into two broad categories: chemical metallurgy and physical metallurgy. Chemical metallurgy is chiefly concerned with the reduction and oxidation of metals, and the chemical performance of metals. Subjects of study in chemical metallurgy include mineral processing, the extraction of metals, thermodynamics, electrochemistry, and chemical degradation (corrosion). In contrast, physical metallurgy focuses on the mechanical properties of metals, the physical properties of metals, and the physical performance of metals. Topics studied in physical metallurgy include crystallography, material characterization, mechanical metallurgy, phase transformations, and failure mechanisms. Modern metallurgists work in both emerging and traditional areas as part of an interdisciplinary team alongside material scientists, and other engineers. Some traditional areas include mineral processing, metal production, heat treatment, failure analysis, and the joining of metals (including welding, brazing, and soldering). Emerging areas for metallurgists include nanotechnology, superconductors, composites, biomedical materials, electronic materials (semiconductors), and surface engineering. Historically, metallurgy has predominately focused on the production of metals. Metal production begins with the processing of ores to extract the metal, and includes the mixture of metals to make alloys.

1. Mineral Processing

Mineral processing manipulates the particle size of solid raw materials to separate valuable materials from materials of no value. Usually, particle sizes must be reduced to efficiently separate valuable materials. Since many size reduction and separation processes involve the use of water, solid-liquid separation processes are part of mineral processing. In order to dissolve an ore in an aqueous solution, it is often necessary to break the large chunks into smaller pieces, thereby increasing the surface area and the rate of dissolution.

2. Extractive Metallurgy

Extractive metallurgy is the practice of separating metals from their ores and refining them into pure metals. To convert a metal oxide or sulfide to a metal, the ore must be reduced either chemically or electrolytically. Various separation techniques are employed to concentrate particles of value and discard waste. In this process, extractive metallurgists are interested in three general streams: the feed, the concentrate (valuable metal oxide or sulfide), and the tailings (waste). Ore bodies often contain more than one valuable metal. Thus the feed might be directly from an ore body, or from a concentrate stream, or even from the tailings of a previous process.

3. Metallurgy in Production Engineering

In production engineering, metallurgy is concerned with the production of metallic components for use in consumer or engineering products. This involves the production of alloys, the shaping, the heat treatment and the surface treatment of the product. Determining the hardness of the metal using the Rockwell, Vickers, and Brinell hardness scales is a commonly used practice that helps better understand the metal's elasticity and plasticity for different applications and production processes. The task of the metallurgist is to achieve balance between material properties such as cost, weight, strength, toughness, hardness, corrosion, fatigue resistance, and performance in temperature extremes.

Common engineering metals are aluminum, chromium, copper, iron, magnesium, nickel, titanium, and zinc. These are most often used as alloys. Much effort has been placed on understanding one very important alloy system, that of purified iron, which has carbon dissolved in it, better known as steel. Cast irons, including ductile iron are also part of this system. Stainless steel or galvanized steel are used where resistance to corrosion is important. Aluminium alloys and magnesium alloys are used for applications where strength and lightness are required. Most engineering metals are stronger than most plastics and are tougher than most ceramics.

The operating environment of the product is very important—a well-designed material will resist expected failure modes such as corrosion, stress concentration, metal fatigue, creep, and environmental stress fracture. Ferrous metals and some aluminium alloys in water and especially in an electrolytic solution such as seawater, corrode quickly. Metals in cold or cryogenic conditions tend to lose their toughness becoming more brittle and prone to cracking. Metals under continual cyclic loading can suffer from metal fatigue. Metals under constant stress in hot conditions can creep.

(Source: New World Encyclopedia Website)

Words and Expressions

sheer a. 纯粹的，完全的，数量大的
unprecedented a. 前所未有的，史无前例的
metallurgy n. 冶金学
intermetallic a. 金属间的
metalworking n. 金属加工，金属制造
mineral n. 矿物
craft n. 工艺，船，飞行器
thermodynamic n. 热动力学
electrochemistry n. 电化学
metallurgist n. 冶金学家
chunk n. 大块，组块
feed n. （机器的）进料装置
ore n. 矿石，矿砂
welding n. 焊接
brazing n. （用锌铜合金）衔接
soldering n. 焊接，钎焊
superconductor n. 超导体
composite n. 复合材料
crystallography n. 晶体学
interdisciplinary a. 多学科的
tailing n. 尾渣
galvanized a. 镀锌的，电镀的
elasticity n. [力] 弹性，弹力，灵活性
strength n. [力] 强度
toughness n. [力] 韧性
hardness n. [力] 硬度
ceramic n. 陶瓷制品，制陶艺术；a. 陶瓷的
fracture n. 断裂，[医] 骨折，（岩层的）裂缝
chemical degradation 化学降解
phase transformation 相转变
raw material 原材料
extractive metallurgy 提炼冶金，提取冶金
mechanical property 力学性能，机械性能
failure mechanism 失效机制
surface area 表面积
biomedical material 生物医用材料

Phrases

coupled with 加上，与……相结合
distinct from 与……不同
rely on 依赖，依靠
be employed to 被用来……
subdivide into 分开，再分成
concern with 关心，专注于
suffer from 遭受，经历，忍受
focus on 关注于，集中于

Notes

(1) Modern civilization started with the discovery of metals. Without metals it would not have been possible to build railway, transport, bridges, buildings, cars, generate power or any of the automobile, electrical and electronic industries.

would not have been possible 译为“几乎不可能的”；electrical 译为“电的，电气的”，而 electronic 译为“电子的”。

参考译文：现代文明始于金属的发现。没有金属，就没有铁路、运输、桥梁、建筑物、汽车、发电或任何汽车、电气和电子工业。

(2) Metallurgy is a domain of materials science and engineering that studies the physical and chemical behavior of metallic elements, their intermetallic compounds, and their mixtures, which are called alloys.

that 引导宾语从句；which 引导定语从句，修饰“mixture”；materials science and engineering 译为“材料科学与工程”。

参考译文：冶金学属于材料科学和工程领域，它研究金属元素，金属间化合物及其称为合金的金属混合物的物理和化学行为。

(3) Extractive metallurgy is the practice of separating metals from their ores and refining them into pure metals. To convert a metal oxide or sulfide to a metal, the ore must be reduced either chemically or electrolytically.

the practice of 译为“……的实践，……的做法”；extractive metallurgy 译为“提取冶金”，它是研究分离和浓缩原材料过程的学科。

参考译文：萃取冶金是将金属从矿石中分离出来并将其提纯为纯金属的过程。为了将金属氧化物或硫化物转化为金属，必须对矿石进行化学或电解还原。

(4) Much effort has been placed on understanding one very important alloy system, that of purified iron, which has carbon dissolved in it, better known as steel.

be placed on 译为“强加于，已处于，受到”；that of 代替了前面的“alloy system”；which 引导定语从句，修饰“purified iron”。

参考译文：人们花了很多精力来了解一种非常重要的合金体系，即溶解了碳的纯铁，也就是我们所熟知的钢。

(5) The operating environment of the product is very important—a well-designed material will resist expected failure modes such as corrosion, stress concentration, metal fatigue, creep, and environmental stress fracture.

破折号后面为一个解释性的分句，用于说明前句的实际意义。

参考译文：产品的操作环境非常重要——精心设计的材料将能抵抗预期的失效方式，比如腐蚀、应力集中、金属疲劳、蠕变和环境应力断裂。

3.2　阅读材料
3.2　Reading Materials

History of Metallurgy

Metallurgy, art and science of extracting metals from their ores and modifying the metals for use. Metallurgy customarily refers to commercial as opposed to laboratory methods. It also concerns the chemical, physical, and atomic properties and structures of metals and the principles whereby metals are combined to form alloys. The present-day use of metals is the culmination of a long path of development extending over approximately 6500 years. It is generally agreed that the first known metals were gold, silver, and copper, which occurred in the native or metallic state, of which the earliest were in all probability nuggets of gold found in the sands and gravels of riverbeds. Such native metals became known and were appreciated for their ornamental and utilitarian values during the latter part of the Stone Age.

1. Earliest Development

Gold can be agglomerated into larger pieces by cold hammering, but native copper cannot, and an essential step toward the Metal Age was the discovery that metals such as copper could be fashioned into shapes by melting and casting in molds; among the earliest known products of this type are copper axes cast in the Balkans in the 4th millennium BCE. Another step was the discovery that metals could be recovered from metal-bearing minerals.

(1) Bronze

In many regions, copper-arsenic alloys, of superior properties to copper in both cast and wrought form, were produced in the next period. This may have been accidental at first. Essentially arsenic-free copper alloys with higher tin content—in other words, true bronze—seem to have appeared between 3000 and 2500 BCE. The discovery of the value of tin may have occurred through the use of stannite, a mixed sulfide of copper, iron, and tin, although this mineral is not as widely available as the principal tin mineral, cassiterite, which must have been the eventual source of the metal.

(2) Iron

An early piece of iron from a trackway in the province of Drenthe, Netherlands, has been dated to 1350 BCE, a date normally taken as the Middle Bronze Age for this area. Certainly, by 1400 BCE in Anatolia, iron was assuming considerable importance, and by 1200-1000 BCE it was being fashioned on quite a large scale into weapons, initially dagger blades. For this reason, 1200 BCE has been taken as the beginning of the Iron Age. By 1000 BCE iron was beginning to be known in central Europe. Its use spread slowly westward. Iron making was fairly widespread in Great Britain at the time of the Roman invasion in 55 BCE. In Asia iron was also known in ancient times, in China by about 700 BCE.

(3) Brass

While some zinc appears in bronzes dating from the Bronze Age, this was almost certainly an accidental inclusion. Brass, as an alloy of copper and zinc without tin, did not appear in Egypt until about 30 BCE, but after this it was rapidly adopted throughout the Roman world, for example, for currency. The general establishment of a brass industry was one of the important metallurgical contributions made by the Romans.

(4) Precious Metals

In addition, by 500 BCE, rich lead-bearing silver mines had opened in Greece. Native gold itself often contained quite considerable quantities of silver.

2. From 500 BCE to 1500 CE

In the thousand years between 500 BCE and 500 CE, a vast number of discoveries of significance to the growth of metallurgy were made. The Greek mathematician and inventor Archimedes, for example, demonstrated that the purity of gold could be measured by determining its weight and the quantity of water displaced upon immersion—that is, by determining its density. In the pre-Christian portion of the period, the first important steel production was started in India, using a process already known to ancient Egyptians. Arsenic, zinc, antimony, and nickel may well have been known from an early date but only in the alloy state. Lead was beaten into sheets and pipes, the pipes being used in early water systems. The metal tin was available and Romans had learned to use it to line food containers. Although the Romans made no extraordinary metallurgical discoveries, they were responsible for, in addition to the establishment of the brass industry, contributing toward improved organization and efficient administration in mining.

Beginning about the 6th century, the most meaningful developments in metallurgy centered on iron making. Great Britain, where iron ore was plentiful, was an important iron-making region. Iron weapons, agricultural implements, domestic articles, and even personal adornments were made. In Spain, another iron-making region, the Catalan forge had been invented, and its use later spread to other areas. The Chinese were the first to realize its advantages. In fact, the Chinese, whose Iron Age began about 500 BCE, appear to have learned to oxidize the carbon from cast iron in order to produce steel or wrought iron indirectly, rather than through the direct method of starting from low-carbon iron.

3. After 1500

During the 16th century, metallurgical knowledge was recorded and made available. Two books were especially influential. One, by the Italian Vannoccio Biringuccio, was entitled De la pirotechnia (Eng. trans., *The Pirotechnia of Vannoccio Biringuccio*, 1943). The other, by the German Georgius Agricola, was entitled *De re metallica*. Biringuccio was essentially a metalworker, and his book dealt with smelting, refining, assay methods and covered metal casting, molding, core making, and the production of such commodities as cannons and cast-iron cannonballs. His book was the first methodical description of foundry practice.

(1) Ferrous Metals

From 1500 to the 20th century, metallurgical development was still largely concerned with improved technology in the manufacture of iron and steel. The most important development of the 19th century was the large-scale production of cheap steel. The first change was the development of the open-hearth furnace by William and Friedrich Siemens in Britain and by Pierre and Émile Martin in France. Another major advance was Henry Bessemer's process, patented in 1855 and first operated in 1856. Soon after the end of the century it replaced wrought iron in virtually every field. Then, with the availability of electric power, electric-arc furnaces were introduced for making special and high-alloy steels.

(2) Light Metals

Another important development of the late 19th century was the separation from their ores, on a substantial scale, of aluminum and magnesium. In the earlier part of the century, several scientists had made small quantities of these light metals, but the most successful was Henri-Étienne Sainte-

Claire Deville, who by 1855 had developed a method by which cryolite, a double fluoride of aluminum and sodium, was reduced by sodium metal to aluminum and sodium fluoride. The process was very expensive, but cost was greatly reduced when the American chemist Hamilton Young Castner developed an electrolytic cell for producing cheaper sodium in 1886.

(3) Welding

One of the most significant changes in the technology of metals fabrication has been the introduction of fusion welding during the 20th century. Before this, the main joining processes were riveting and forge welding. Both had limitations of scale, although they could be used to erect substantial structures. In 1895 Henry-Louis Le Chatelier stated that the temperature in an oxyacetylene flame was 3500℃ (6300℉), some 1000℃ higher than the oxyhydrogen flame already in use on a small scale for brazing and welding. The first practical oxyacetylene torch, drawing acetylene from cylinders containing acetylene dissolved in acetone, was produced in 1901. With the availability of oxygen at even lower cost, oxygen cutting and oxyacetylene welding became established procedures for the fabrication of structural steel components. The metal in a join can also be melted by an electric arc, and a process using a carbon as a negative electrode and the workpiece as a positive first became of commercial interest about 1902.

(4) Metallography

The 20th century has seen metallurgy change progressively, from an art or craft to a scientific discipline and then to part of the wider discipline of materials science. In extractive metallurgy, there has been the application of chemical thermodynamics, kinetics, and chemical engineering, which has enabled a better understanding, control, and improvement of existing processes and the generation of new ones.

This greater scientific understanding has come largely from a continuous improvement in microscopic techniques for metallography, the examination of metal structure. The first true metallographer was Henry Clifton Sorby of Sheffield, England, who in the 1860s applied light microscopy to the polished surfaces of materials such as rocks and meteorites. Sorby eventually succeeded in making photomicrographic records, and by 1885 the value of metallography was appreciated throughout Europe, with particular attention being paid to the structure of steel.

(Source: Paul G Shewmon, *Metallurgy*. Britannica Website)

Words and Expressions

principle *n.* 准则，原理，定律
stannite *n.* [化] 亚锡酸盐
cassiterite *n.* [矿] 锡石
furnace *n.* 火炉，熔炉
cryolite *n.* 冰晶石
fluoride *n.* 氟化物
rivet *n.* 铆钉；*vt.* 铆接，固定
cylinder *n.* 圆柱体，圆筒，汽缸
meteorite *n.* [地] 陨石
fabrication *n.* 制造，生产，组装
oxidize *vt.* 使氧化；使生锈
cold hammering 冷锻，冷压
lead-bearing 含铅的

3.3　问题与讨论
3.3　Questions and Discussions

(1) What is metallurgy?

(2) The categories of metallurgy and its difference?

(3) What are the emerging areas for metallurgy scientists?

(4) What is extractive metallurgy?

(5) What is the task for metallurgist?

(6) Introduce expected failure modes for metallic materials.

单元4 金属晶体结构
Unit 4 Crystal Structure of Metals

扫码查看
讲课视频

4.1 教学内容
4.1 Teaching Materials

1. The Structure of Metals

Metals are usually crystalline when in the solid form. While very large single crystals can be prepared, the normal metallic object consists of an aggregate of many very small crystals. Metals are therefore polycrystalline. The crystals in these materials are normally referred to as its grains. There is the basic structure inside the grains themselves: that is, the atomic arrangements inside the crystals. This form of structure is logically called the crystal structure. Fortunately, most metals crystallize in one of three relatively simple structures: the face-centered cubic, the body-centered cubic, and the close-packed hexagonal.

2. Unit Cells

The unit cell of a crystal structure is the smallest group of atoms possessing the symmetry of the crystal which, when repeated in all directions, will develop the crystal lattice. Figure 1(a) shows the unit cell of the body-centered cubic lattice. Eight unit cells are combined in Figure 1(b) in order to show how the unit cell fits into the complete lattice. Note that atom a of Figure 1(b) does not belong uniquely to one unit cell, but is a part of all eight unit cells that surround it. Therefore, in the interior of a crystal, each corner atom of a unit cell is equivalent to atom a of Figure 1(b) and contributes one-eighth of an atom to a unit cell. In addition, each cell also possesses an atom located at its center that is not shared with other unit cells. The body-centered cubic lattice thus has two atoms per unit cell, as shown in Figure 1(c).

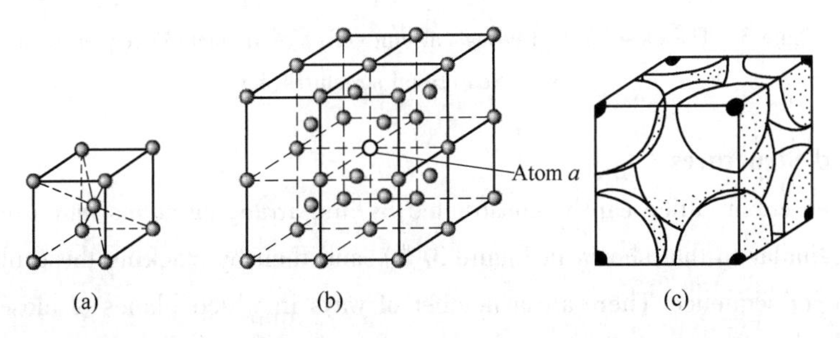

Figure 1 Body-centered cubic unit cell(a), eight unit cells of the body-centered cubic lattice(b), and cut view of a unit cell(c)

The unit cell of the face-centered cubic lattice is shown in Figure 2(a). In this case, the unit cell has an atom in the center of each face. The eight corner atoms again contribute one atom to the cell, as shown in Figure 2(b). There are also six face-centered atoms to be considered, each a part of two-unit cells. These contribute six times one-half an atom, or three atoms. The face-centered cubic lattice has a total of four atoms per unit cell, or twice as many as the body-centered cubic lattice.

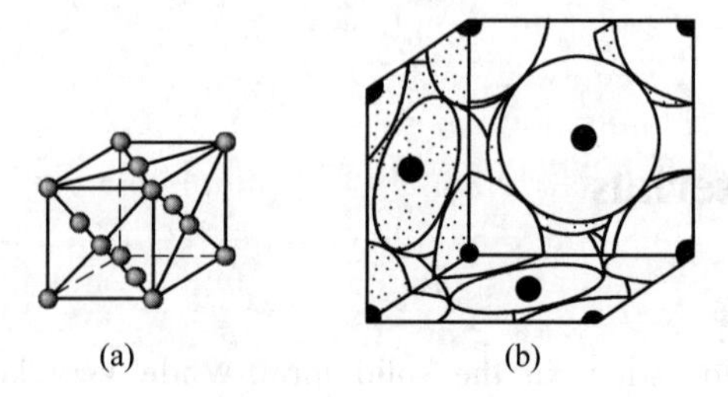

Figure 2 Face-centered cubic unit cell(a) and cut view of a unit cell(b)

The configuration of atoms most frequently used to represent the hexagonal close packed structure is shown in Figure 3. This group of atoms contains more than the minimum number of atoms needed to form an elementary building block for the lattice; therefore it is not a true unit cell. However, because the arrangement of Figure 3(a) brings out important crystallographic features, including the sixfold symmetry of the lattice, it is commonly used as the unit cell of the close-packed hexagonal structure.

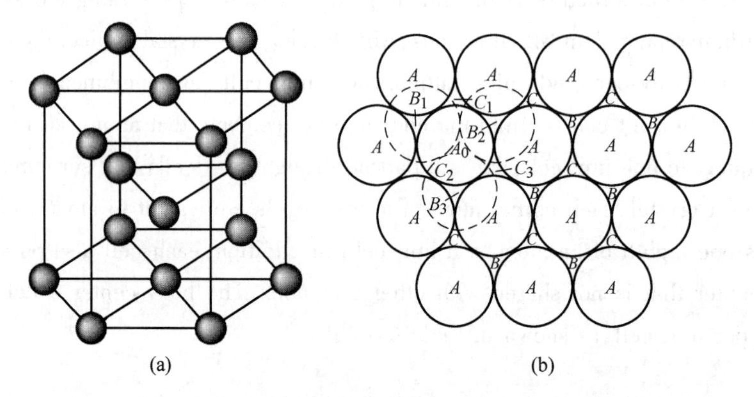

Figure 3 The close-packed hexagonal unit cell(a) and stacking sequences in close-packed crystal structures(b)

3. Close-Packed Structures

The face-centered cubic lattice can be constructed by first arranging atoms into a number of close-packed planes, similar to that shown in Figure 3(b), and then by stacking these planes over each other in the proper sequence. There are a number of ways in which planes of closest packing can be stacked. One sequence gives the close-packed hexagonal lattice, another the face-centered cubic lattice. For example, consider the close-packed plane of atoms in Figure 3(b). The face-centered cubic stacking order is *ABCABCABC*. In the close-packed hexagonal structure, the atoms in every other plane fall directly over one another, corresponding to the stacking order ABABAB…

There is no basic difference in the packing obtained by the stacking of spheres in the face-centered cubic or the close-packed hexagonal arrangement, since both give an ideal close-packed structure. There is, however, a marked difference between the physical properties of hexagonal close-packed metals (such as cadmium, zinc, and magnesium) and the face-centered cubic metals, (such as aluminum, copper, and nickel), which is related directly to the difference in their crystalline structure.

4. Crystal Structures of the Metallic Elements

Some of the most important metals are classified according to their crystal structures in Table 1. A number of metals are polymorphic, that is, they crystallize in more than one structure. The most important of these is iron, which crystallizes as either body-centered cubic or face-centered cubic, with each structure stable in separate temperature ranges.

Table 1 Crystal structure of some of the most important metallic elements

Face-Centered Cubic	Closed-Packed Hexagonal	Body-Centered Cubic
Iron(911.5 to 1396℃)	Magneium	Iron(below 911.5 and from 1396 to 1538℃)
Copper	Zinc	Titanium(882 to 1670℃)
Silver	Titanium(below 882℃)	Zirconium(863 to 1855℃)
Gold	Zirconium(below 863℃)	Tungsten
Aluminum	Beryllium	Vanadium
Nickel	Cadmium	Molybdenum
Lead		Alkali Metals(Li, Na, K, Rb, Ca)
Platinum		

(Source: Reza Abbaschian, et al. *Physical Metallurgy Principles*. Stamford: Gengage Learning, 2009)

Words and Expressions

optical *a.* 光学的
magnification *n.* 放大倍数
polycrystalline *n.* 多晶材料
grain *n.* 晶粒，颗粒，细粒
octahedral *a.* 八面体的
configuration *n.* 布局，构造，配置
polymorphic *n.* 多态的
naked eye 肉眼
atomic arrangement 原子排列
crystal lattice 晶格，晶体点阵
unit cell 单位晶胞
body-centered cubic lattice 体心立方点阵
face-centered cubic lattice 面心立方点阵
hexagonal close-packed structure 密排六方结构

Phrases

consist of 由……组成
associated with 与……有关系
fall into 落入，分成

the order of	数量级
be equivalent to	相似于，等价于
share with	分与，分配，和……分享
bring out	出版，生产，使显示，公布

Notes

(1) Because of their very small sizes, an optical microscope, operating at magnifications between about 100 and 1000 times, is usually used to examine the structural features associated with the grains in a metal.

associated with 译为“与……有关系的”；optical microscope 译为“光学显微镜”。

参考译文：由于它们的尺寸非常小，通常在光学显微镜下放大100到1000倍来观察金属晶粒的结构特征。

(2) The unit cell of a crystal structure is the smallest group of atoms possessing the symmetry of the crystal which, when repeated in all directions, will develop the crystal lattice.

unit cell 译为“单位晶胞”；when repeated in all directions 作为插入语；crustal lattice 译为“晶格”。

参考译文：晶胞是具有晶体对称性的最小的一组原子群，当它们在各个方向重复时，就会形成晶格。

(3) The face-centered cubic lattice can be constructed by first arranging atoms into a number of close-packed planes, similar to that shown in Figure 3(b), and then by stacking these planes over each other in the proper sequence.

in the proper sequence 译为“以适当的顺序”；first…and then…连接两个并列句子。

参考译文：面心立方晶格可以通过如下方法构建：首先将原子排列成若干紧密排列的平面［见图3(b)］，然后将这些平面按适当的顺序相互堆叠。

(4) There is no basic difference in the packing obtained by the stacking of spheres in the face-centered cubic or the close-packed hexagonal arrangement, since both give an ideal close-packed structure.

no basic difference in 译为“没有什么不同”;face-centered cubic(面心立方)与 close-packed”（密排立方）为金属常见的两种密排结构。

参考译文：面心立方和密排六方排列的球堆积没有本质的区别，两者都是理想的密排结构。

4.2 阅读材料

4.2 Reading Materials

Nanocrystalline and Amorphous Metals

1. Nanocrystalline Metals

In a very real sense, the modern aerospace industry is built on a foundation of nanocrystalline metals. Age-hardenable aluminum alloys demonstrate manipulation and control of microstructural features at length scales of 1 ~ 10nm, producing very high levels of strength. Ni-based superalloys demonstrate control of features from 10 ~ 1000nm in alloys that retain a significant fraction of their room temperature strength at a maximum operating temperature that is 85% of the absolute melting temperature of the alloy. In steels, dimensions in pearlite colonies are controlled over length scales of 100 ~ 500nm, producing a remarkable balance of exceptional strength and fracture resistance that no other structural material can match.

It is well known that dramatic strengthening can be produced by reducing the relevant length scales for deformation in metallic alloys. Three length scales are shown in Figure 1. Reduction of the grain diameter (D) produces the well-known Hall-Petch relationship, while reducing the diameter(δ) of strengthening precipitates or dispersoids at a given volume fraction increases the resistance to dislocation motion through bowing of dislocations around the barriers (Orowan strengthening). Below some critical diameter, particles are cut by moving dislocations, and the resistance for flow decreases with further decrease in δ. Finally, increasing the volume fraction of the strengthening precipitates at a fixed particle size decreases their mean separation (d), producing a third strengthening contribution. A great deal of effort is evident in the literature to produce nanocrystalline microstructures that approach these values. Processing of high quality, fully dense nanocrystalline metals provides a significant technological challenge. Both "bottom-up" (where nanocrystalline powders are produced and consolidated) and "top-down" (where a bulk material is processed to produce nanocrystalline microstructural dimensions through work hardening or similar methods) approaches are being pursued. Since strengthening increases non-linearly with decreasing microstructural dimensions, significant strengthening is expected as these techniques are refined.

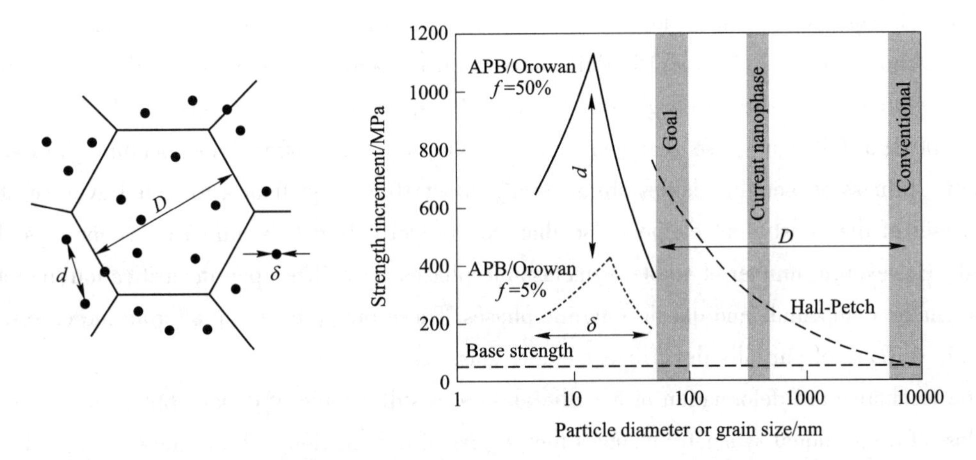

Figure 1 Influence of length scale on strengthening in nanoscale metals
(The three length scales, D, d and δ influence different strengthening mechanisms and can provide additive strengthening contributions)

2. Amorphous Metals

Amorphous metals are a relatively new class of materials, which possess exceptional specific strength along with other functional properties including exceptional "soft" magnetic properties, exceptional corrosion resistance and unusual damping behavior. The stability and mechanical properties of partially devitrified alloys also provide surprising responses that are not yet understood. Thus, the research on amorphous metals for structural applications is of a more fundamental and exploratory nature relative to other metals technologies presented here. Expanded applications are expected in the coming years as the fundamental behaviors of metallic glasses become better established.

Of primary importance is the stability of metallic glasses. Most glasses require very rapid quenching from the liquid to retain the amorphous structure in the solid state, so that material with at least one dimension of the solid product is of a length scale less than 100μm. Consolidation of such material is difficult, since most amorphous metals crystallize below temperatures typically used for consolidation. On the other hand, a very small number of metallic glasses can be produced in the fully amorphous condition with cooling rates as slow as a few tenths of a degree Celsius per second. These bulk metallic glasses (BMGs) are favored for structural applications, and so the basic features that lead to this exceptional stability need to be established. Recent results have established a dominant role of the relative size and number of atoms (topology). A distinct system topology has been identified for BMGs, where the solvent atom is the largest atomic constituent, the next most concentrated constituent is the smallest solute, and constituents of intermediate size have the lowest concentrations. More recently, the principle of efficient atomic packing has been established, and this has led to the discovery that specific atomic radius ratios (relative to the solvent atom radius) are preferred in metallic glasses.

Metallic glasses typically have a fracture stress that is about 2% of the elastic modulus, so that metallic glasses have roughly double the strength of crystalline alloys of the same alloy base element. Thus, strengths of up to 1500MPa have been measured for Al alloys, and strengths up to 4GPa are obtained for Fe-based glasses. The specific strengths of amorphous metals are typically 400~500MPa/(Mg · m^3), so that compelling opportunities are offered by metallic glasses. The specific stiffness of metallic glasses are generally about 10% lower than a weighted average of the stiffnesses of the constituent elements, so that this property is not as attractive as for crystalline metals. However, a number of stable or metastable phases crystallize upon devitrification, including intermetallic compounds and quasicrystalline phases. These precipitates can offer an increase in the specific stiffness of partially devitrified metallic glasses.

The mechanism of deformation of metallic glasses is still a topic of debate, but it is certain that the loss of translational symmetry changes the physics of deformation and plasticity. This produces a unique plastic response in metallic glasses. Amorphous metals deform by intensely localized shear within a narrow band, typically on the order of a few tens of nanometers thick. Within the bands, strains over 100% can be achieved, but the number of these bands is very small, so that there is typically no macroscopic ductility measured in unconstrained loading, such as in a tensile test. On the other hand, constrained deformation leads to extensive slip multiplicity and plasticity, so that amorphous metals can be effectively rolled and formed below the glass transition temperature. The intrinsic resistance to fracture is generally good for amorphous metals, and K_{I_c} values are typically of the order of 20MPa/m. Thus, the most serious limitation of metallic glasses for fracture-critical applications is the absence of general plasticity. In-situ composites of cocontinuous BMG and ductile crystalline metal phases have been produced, and these provide enhanced plasticity and fracture properties. Unique nanocrystalline microstructures can be produced by controlled crystallization of the fully amorphous product, including nanocrystalline precipitates homogeneously distributed in an amorphous alloy matrix. In some systems, both the strength and ductility increase in this partially crystalline state. Other alloys produce nanocrystalline intermetallic or quasicrystalline precipitates, providing a credible path for increasing the specific stiffness.

3. Quantitative Description of Microstructures

Models based on simple scalar descriptions of microstructure, such as Hall-Petch strengthening (mean grain size), Orowan strengthening (mean interparticle spacing), and rule-of-mixtures strengthening (mean reinforcement volume fraction), adequately predict macroscopic strength or stiffness in a range of metallic materials. However, models for predicting fracture properties (such as ductility and toughness) of these same materials are inadequate, and often do not provide even a correct representation of trends. A fundamental distinction is that strength and stiffness are often controlled by the mean values of relevant microstructural features, and fracture properties are controlled by extremes in the distributions (that is, the "tails" of distributions) in relevant microstructural features.

In discontinuously reinforced metals, a quantitative microstructural description is required as the basis from which physically based models of ductility, fracture toughness and damage tolerance may be developed. The characteristics of discontinuously reinforced metals that emphasize the importance of the second phase distribution result from the extreme differences in elastic and plastic response between the matrix and reinforcement. Dramatic differences in elastic constants, thermal expansion, strength and plasticity lead to residual stresses, stress concentrations, and elastic and plastic constraint. The spatial distribution of the particulates controls the local magnitude of these features, so that distribution becomes a controlling consideration. Further, interaction between local events related to particle distribution, such as clustering, and interaction with other microstructural features, such as oxide inclusions, provide a relevant system for interrogating and establishing multi-scale effects.

These same comments apply to the development and accelerated implementation of many conventional metal alloys, including alloys of Al, Ti and Ni, and advanced metallic materials. Quantified three-dimensional (3D) descriptions of microstructural features, including the size, shape, distribution, orientation and volume fraction of grains, pores, and intrinsic and extrinsic flaws are required. Rigorous and quantifiable definitions of microstructural concepts such as homogeneity and clustering are also necessary.

(Source: Oleg N Senkov, et al. *Metallic Materials with High Structural Efficiency*. Dordrecht: Kluwer Academic Publishers, 2004)

Words and Expressions

nanocrystalline　*n.* 纳米晶
amorphous　*a.* 无定形的，非晶形的
slip　*n.* 滑移
damping　*n.* ［物］衰减，减幅，阻尼
devitrify　*v.*（玻璃）变硬，析晶，使不透明
quasicrystalline　*n.* 准晶体
microstructure　*n.* 微观结构
deformation　*n.* 变形
intrinsic　*a.* 内在的，固有的
homogeneity　*n.* 同质，同种
magnitude　*n.* 震级，大小，数量（级）
Ni-based superalloy　镍基高温合金
non-linearly　非线性
atom radius　原子半径
fracture stress　断裂应力
elastic modulus　弹性模量
extrinsic flaw　外部缺陷

4.3　问题与讨论

4.3　Questions and Discussions

(1) Discuss the basic three structures of metals.
(2) What is unit cell?
(3) For fcc, how many atoms in a unit cell?
(4) The stacking differences for fcc and hcp.
(5) Why there are different crystalline structures for the same metal?

单元 5　晶体结构缺陷
Unit 5　Imperfections of the Crystal Structure

扫码查看
讲课视频

5.1　教学内容
5.1　Teaching Materials

1. Dislocations

Dislocations are line defects of the crystal lattice, which border the regions in the crystal interior in which slip has occurred (Figure 1). The measure for the quantity of dislocations in the crystal is given as the dislocation density—the total length of all of the dislocation lines per unit volume. The main characteristic of each dislocation is the Burgers vector, which describes the magnitude and direction of slip movement associated with the dislocation. The classification of dislocations is based on the mutual orientation of the dislocation line and the Burgers vector of the dislocation. The direction of an edge dislocation is perpendicular to the Burgers vector. In contrast, the direction of a screw dislocation line is parallel to the Burgers vector. Most of the dislocations are mixed dislocations. They are a combination of screw and edge segments. Mixed dislocations usually form dislocation loops.

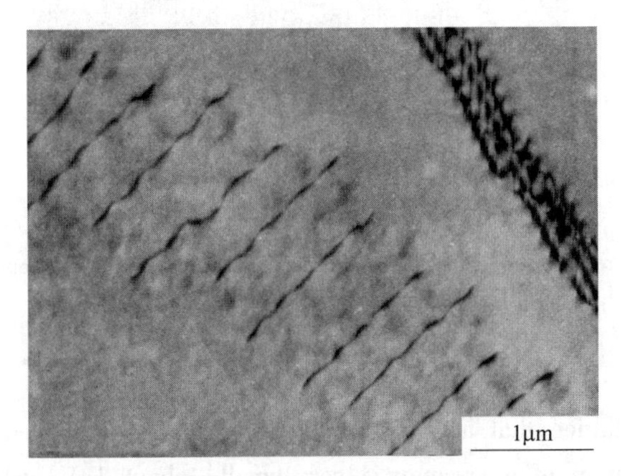

Figure 1　Real crystals contain a significant number of dislocations
(Quenched stainless austenitic steel)

When the magnitude of the Burgers vector equals a whole lattice vector, it is referred to as a unit or perfect dislocation. Such dislocations, when passing through the crystal, do not change the arrangement of atoms in the lattice because a complete lattice translation occurs. Dislocations with a Burgers vector not equal to a whole lattice vector are referred to as imperfect or, more often, as partial dislocations. A basic physical characteristic of metals and alloys that governs the mode of dislocations movement and the type of dislocation reactions is the stacking fault energy (SFE)—the energy necessary to produce a unit area of stacking fault (SF) in a perfect crystal. The stacking faults are the simplest type of planar defects.

2. Multiplication of Dislocations

The number of dislocations in a crystal is changed by mechanical processing. The density of dislocations in a well-annealed crystal varies from $10^6 cm^{-2}$ to $10^8 cm^{-2}$. After 30% to 40% cold plastic deformation, the dislocation density increases to $10^{11} \sim 10^{12} cm^{-2}$. One of the main sources for the multiplication of dislocations during plastic deformation is a Frank-Read source (Figure 2). A Frank-Read source is a region of the crystal with a high density of defects, which is capable of generating dislocations when the shear stress reaches a certain critical level. The process is repeated when a new dislocation breaks out and starts moving away from the source. The source continues to emit dislocations of the same type and can generate an unlimited number of dislocations if the applied stress remains in excess of the critical value. The dislocations moving through the crystal have to overcome numerous obstacles. If the obstacle is difficult to overcome—for example, a grain boundary, an inclusion interface, or a field of stress concentration—the dislocations stop and form dislocation pileups.

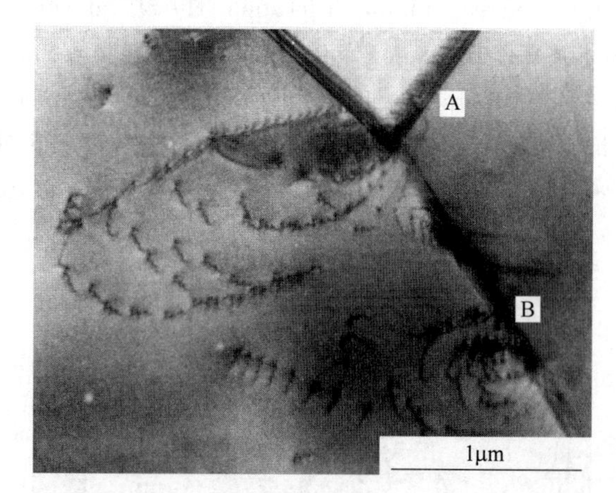

Figure 2 Frequently, grain boundaries contain Frank-Read sources
(Quenched austenitic nitrogen steel Fe-18Cr-14Mn-0,6N)

3. Vacancies

The sites in a crystal lattice that are not occupied by atoms are called vacancies. The energy necessary for the formation of a vacancy is very small—about 1eV—which explains the large concentration of these point defects under thermodynamic equilibrium conditions in metals and alloys. Sources for the formation of vacancies are the free surfaces of the crystals and the internal defects (dislocations, grain and subgrain boundaries, phase interfaces). Vacancies in excess of their equilibrium concentration are generated most often during quenching from high temperatures, during plastic deformation, during ion bombardment, during bombardment by high-energy nuclear particles, or, in some intermetallic compounds, as a result of stoichiometric deviations. They are also produced by the oxidation of some metals, such as Mg, Ni, Cu, Zn, and Cd. Vacancies play a significant part in nonconservative dislocation movement(climb) and in the processes involving a diffusion transport of atoms—for example, polygonization and recovery. They thus play an important role during the processing of metals and alloys by plastic deformation at elevated temperatures.

4. Grain and Sub-Grain Boundaries

The boundaries of grains belong to the group of two-dimensional lattice imperfections. They are interfaces that separate regions in the interior of the material at which the crystal lattice changes orientation. The type and nature of the boundaries depend on the misorientation angle of the two adjoining grains and on the orientation of the interface boundary plane to them. In a low-angle boundary, the angle does not exceed several degrees. When the angle exceeds 10 to 15 degrees, the boundary is known as a random high-angle boundary. This boundary of several interatomic distances in width is in a high state of disorder compared to the matrix crystal structure—the atoms are out of their normal positions, the interatomic bonds are distorted, and consequently the boundary is associated with a higher energy.

5. Twins

A twin is a part of the crystal that has a crystal lattice identical to that of the base crystal(matrix), but with a crystallographic orientation that is a mirror image of the matrix orientation. Twins are always bordered by a pair of parallel coherent boundaries at which is realized the symmetrical tilt between the two twin-related crystals. Because of their extremely low energy due to the perfect fitting of boundary atoms into the lattices of both grains, the coherent boundaries are regarded as special high-angle grain boundaries.

(Source: Ganka Zlateva, al et. *Microstructure of Metals and Alloys*. Boca Rato: CRC Press, 2008)

Words and Expressions

imperfection *n.* 缺陷
dislocation *n.* 位错，[医] 脱臼
bombardment *n.* 轰炸；炮击
interface *n.* 界面
vacancy *n.* 空位
inclusion *n.* 夹杂物，杂质
polygonization *n.* 多边化
interatomic *a.* 原子间的
misorientation *n.* 定向误差，取向误差
crystallographic *a.* 结晶的；结晶（学）的
twin *n.* 孪晶
line defect 线缺陷，位错
Burgers vector 伯格斯矢量
edge dislocation 刃型位错
screw dislocation 螺型位错
mixed dislocation 混合位错
perfect dislocation 全位错
partial dislocation 不全位错
stacking fault energy 位错能
Frank-Read source 弗兰克-里德位错源
grain boundary 晶界
dislocation pileup 位错堆积
point defect 点缺陷
subgrain boundary 亚晶界
planar defect 面缺陷
shear stress 剪应力
coherent boundary 共格晶界
intermetallic compound 金属间化合物

Phrases

pass through	通过，穿过
in contrast	对比，相比而言
perpendicular to	垂直于
parallel to	平行于
in excess of	超过
play an important role	起了重要的作用
identical to	等同于，相同

Notes

(1) The direction of an edge dislocation is perpendicular to the Burgers vector. In contrast, the direction of a screw dislocation line is parallel to the Burgers vector. Most of the dislocations are mixed dislocations. They are a combination of screw and edge segments. Mixed dislocations usually form dislocation loops.

perpendicular to 译为“垂直于……”；parallel to 译为“平行于……”；edge segments（刃位错）、screw dislacation（螺位错）和 mixed dislocations（混合位错）是位错的主要存在形式。

参考译文：刃位错的方向垂直于伯格斯矢量；相反，螺位错方向平行于伯格斯矢量；大多数位错是混合位错，它是螺位错和刃位错的结合。混合型位错通常形成位错环。

(2) A basic physical characteristic of metals and alloys that governs the mode of dislocations movement and the type of dislocation reactions is the stacking fault energy (SFE) — the energy necessary to produce a unit area of stacking fault (SF) in a perfect crystal.

that 引导定语从句；破折号后为宾语补定语，解释 the stacking fault energy 的含义。

参考译文：决定金属和合金位错运动方式和位错反应类型一个基本物理特性是其层错能（SFE），即在一个理想晶体中产生单位面积的层错（SF）所需的能量。

(3) A Frank-Read source is a region of the crystal with a high density of defects, which is capable of generating dislocations when the shear stress reaches a certain critical level.

which 引导定语从句，when 引导条件状语从句；capable of 译为“能够……”。

参考译文：弗兰克-里德位错源是晶体中缺陷密度高的区域，当切应力达到一定的临界水平时，该区域能够产生位错。

(4) Vacancies in excess of their equilibrium concentration are generated most often during quenching from high temperatures, during plastic deformation, during ion bombardment, during bombardment by high-energy nuclear particles, or, in some intermetallic compounds, as a result of stoichiometric deviations.

inexcess of 译为“超过”；during 连接了三个并列结构；intermetallic compounds 译为“金属间化合物”；as a result of 译为“由于……结果”。

参考译文：超过平衡浓度的空位通常是在高温淬火，塑性变形，离子轰击，高能核粒子轰击或某些金属间化合物由于化学计量偏差而产生的。

(5) Twins are always bordered by a pair of parallel coherent boundaries at which is realized

the symmetrical tilt between the two twin-related crystals.

twins 译为 “孪晶”; coherent boundaries 译为 “共格晶界”; at which = where，引导定语从句。

参考译文：孪晶总是以一对平行的共格晶界为边界，在该边界处通过两个孪生晶体间的对称倾斜而实现。

5.2　阅读材料

5.2　Reading Materials

Movement of Dislocations

1. Concept of Slip

There are two basic types of dislocation movement. Glide or conservative motion occurs when the dislocation moves in the surface which contains both its line and Burgers vector: a dislocation able to move in this way is glissile, one which cannot is sessile. Climb or non-conservative motion occurs when the dislocation moves out of the glide surface, and thus normal to the Burgers vector. Glide of many dislocations results in slip, which is the most common manifestation of plastic deformation in crystalline solids. It can be envisaged as sliding or successive displacement of one plane of atoms over another on so-called slip planes. Discrete blocks of crystal between two slip planes remain undistorted as illustrated in Figure 1.

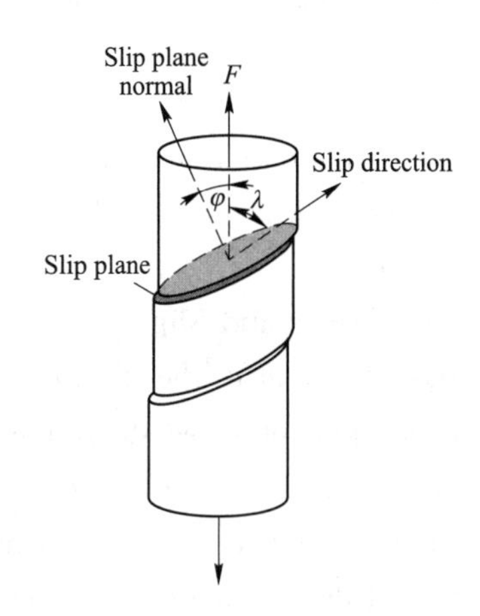

Figure 1　Illustration of the geometry of slip in a cylindrical crystal. Note that ($\varphi+\lambda$) not equals to 90 in general

The slip planes and slip directions in a crystal have specific crystallographic forms. The slip planes are normally the planes with the highest density of atoms, i. e. those which are most widely spaced, and the direction of slip is the direction in the slip plane corresponding to one of the shortest lattice translation vectors. Often, this direction is one in which the atoms are most closely spaced. The shortest lattice vectors are $\frac{1}{2}\langle 111\rangle$; $\frac{1}{2}\langle 110\rangle$ and $\frac{1}{3}\langle 1120\rangle$ in the body-centered cubic, face-centered cubic and close-packed hexagonal systems, respectively. Thus, in close-packed hexagonal crystals, slip often occurs on the (0001) basal plane in directions of the type $\langle 1120\rangle$ and in face-centered cubic metals on {111} planes in $\langle 110\rangle$ directions. In body-centered cubic metals the slip direction is the $\langle 111\rangle$ close-packed direction, but the slip plane is not well defined on a macroscopic scale. Microscopic evidence suggests that slip occurs on {112} and {110} planes and that {110} slip is preferred at low temperatures. A slip plane and a slip direction in the plane constitute a slip system. Face-centered cubic crystals have four {111} planes with three $\langle 110\rangle$ directions in each, and therefore have twelve {111}

<110> slip systems.

Slip results in the formation of steps on the surface of the crystal. These are readily detected if the surface is carefully polished before plastic deformation. A characteristic shear stress is required for slip. Consider the crystal illustrated in Figure 1 which is being deformed in tension by an applied force F along the cylindrical axis. If the cross-sectional area is A the tensile stress parallel to F is $\sigma = \frac{F}{A}$. The force has a component $F\cos\lambda$ in the slip direction, where λ is the angle between F and the slip direction. This force acts over the slip surface which has an area $\frac{A}{\cos\varphi}$, where φ is the angle between F and the normal to the slip plane. Thus the shear stress τ, resolved on the slip plane in the slip direction, is:

$$\tau = \frac{F}{A}\cos\varphi\cos\lambda \tag{1}$$

If F_c is the tensile force required to start slip, the corresponding value of the shear stress τ_c is called the critical resolved shear stress (CRSS) for slip. The quantity $\cos\varphi\cos\lambda$ is known as the Schmid factor.

2. Dislocations and Slip

It was shown that the theoretical shear stress for slip was many times greater than the experimentally observed stress, i. e. τ_c. The low value of τ_c can be accounted for by the movement of dislocations. Consider the edge dislocation, this could be formed in a different way to that described as follows: cut a slot along *AEFD* in the crystal shown in Figure 2 and displace the top surface of the cut *AEFD* one lattice spacing over the bottom surface in the direction *AB*. An extra half-plane *EFGH* and a dislocation line *FE* are formed. Apart from the immediate region around the dislocation core *FE*, the atoms across *AEFD* are in perfect registry. Only a relatively small applied stress is required to move the dislocation along the plane *ABCD* of the crystal in the way demonstrated in Figure 3. This can be understood from the following argument. Well away from the dislocation, the atom spacings are close to the perfect crystal values, and a shear stress as high as the theoretical value would be required to slide them all past each other at the same time. Near the dislocation line itself, some atom spacings are far from the ideal values, and small relative changes in position of only a few atoms are required for the dislocation to move. For example, a small shift

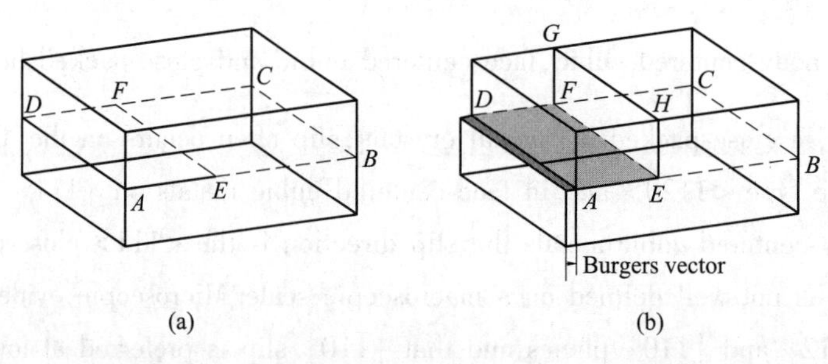

Figure 2 Formation of a pure edge dislocation *FE*
(a) Before; (b) After slide

of atom 1 relative to atoms 2 and 3 in Figure 3(a) effectively moves the extra half-plane from x to y [Figure 3(b)], and this process is repeated as the dislocation continues to glide. The applied stress required to overcome the lattice resistance to the movement of the dislocation is the Peierls Nabarro stress and is much smaller than the theoretical shear stress of a perfect crystal. Figure 3 demonstrates why the Burgers vector is the most important parameter of a dislocation. The glide of one dislocation across the slip plane to the surface of the crystal produces a surface step equal to the Burgers vector.

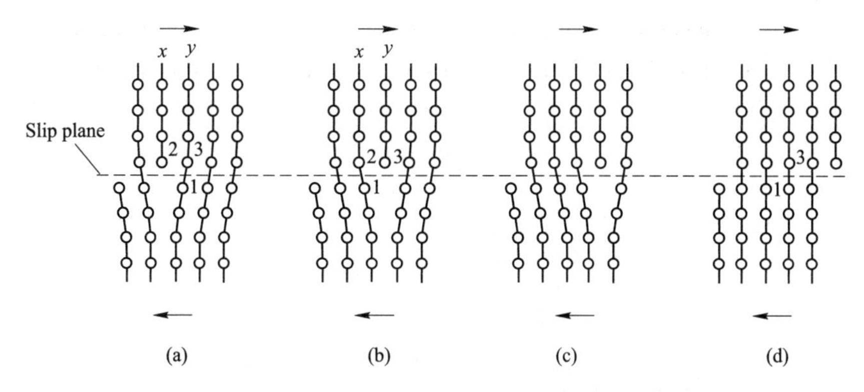

Figure 3 Movement of an edge dislocation: the arrows indicate the applied shear stress

3. The Slip Plane

In Figure 2 the edge dislocation has moved in the plane *ABCD* which is the slip plane. This is uniquely defined as the plane which contains both the line and the Burgers vector of the dislocation. The glide of an edge dislocation is limited, therefore, to a specific plane. The movement of a screw dislocation, for example from *AA′* to *BB′* in Figure 4, can also be envisaged to take place in a slip plane, i. e. *LMNO*, and a slip step is formed. However, the line of the screw dislocation and the Burgers vector do not define a unique plane and the glide of the dislocation is not restricted to a specific plane. It will be noted that the displacement of atoms and hence the slip step associated with the movement of a screw dislocation is parallel to the dislocation line, for that is the direction of its Burgers vector. This can be demonstrated further by considering a plan view of the atoms above and below a slip plane containing a screw dislocation (Figure 5).

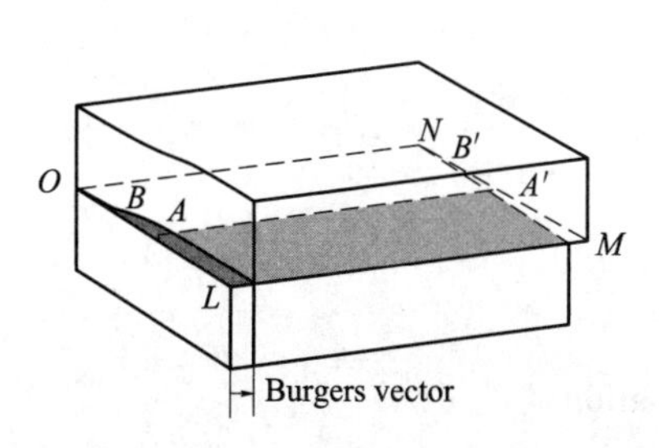

Figure 4 Formation and then movement of a pure screw dislocation *AA′* to *BB′* by slip

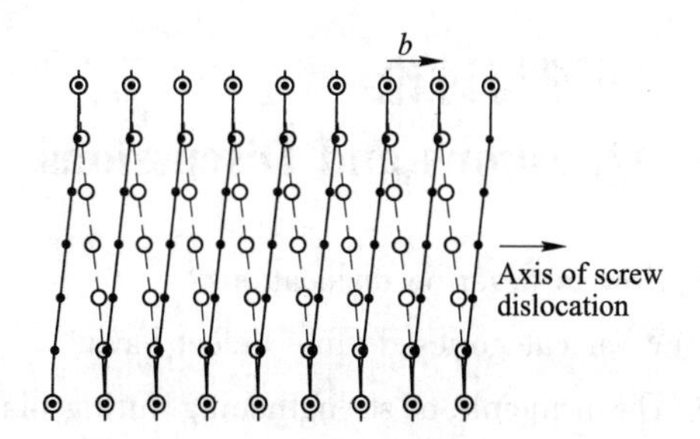

Figure 5 Arrangement of atoms around a screw dislocation [Open circles above plane of diagram, filled circles below (for right-handed screw)]

In the examples illustrated in Figures 2 and 4 it has been assumed that the moving dislocations remain straight. However, dislocations are generally bent and irregular, particularly after plastic deformation, as can be seen in the electron micrographs. A more general shape of a dislocation is shown in Fignre 6(a). The boundary separating the slipped and unslipped regions of the crystal is curved, i. e. the dislocation is curved, but the Burgers vector is the same all along its length. It follows that at point E the dislocation line is normal to the vector and is therefore pure edge and at S is parallel to the vector and is pure screw. The remainder of the dislocation(M) has a mixed edge and screw character. The Burgers vector b of a mixed dislocation, XY in Figure 6(b), can be resolved into two components by regarding the dislocation as two coincident dislocations; a pure edge with vector b_1 of length $b \sin\theta$ at right angles to XY, and a pure screw with vector b_2 of length $b \cos\theta$ parallel to XY:

$$b_1 + b_2 = b \tag{2}$$

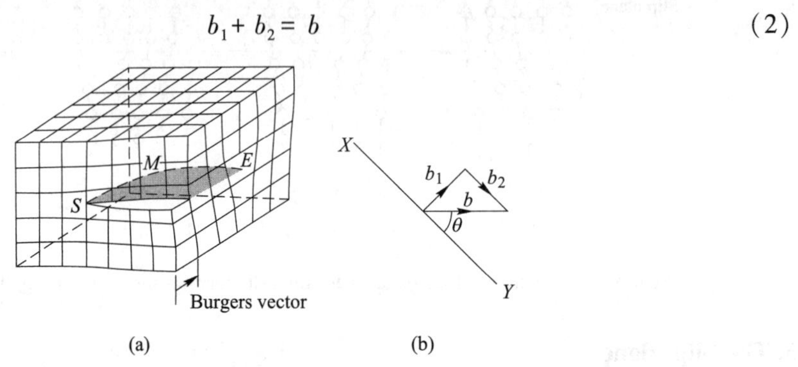

Figure 6　Mixed dislocations

(a) The curved dislocation *SME* is pure edge at E and pure screw at S;

(b) Burgers vector b of dislocation XY is resolved into a pure edge component b_1 and a pure screw component b_2

(Source: D Hull, D J Bacon. *Introduction to Dislocations*. Oxford: Elsevier Butterworth-Heinemann, 2001)

Words and Expressions

glide　*n.* 滑移（slip）

climb　*n.* 攀移

cross-section　横断面，截面

conservative motion　守恒运动

slip plane（direction）　滑移面（线）

5.3　问题与讨论

5.3　Questions and Discussions

(1) How to describe dislocations?

(2) What categories do line defect have?

(3) The principle of strengthening during plastic deformation.

(4) What will happen during quenching?

(5) klow to distinguish low-angle boundary and high-angle boundary?

(6) What is twin boundary?

第二部分　金属合金

Part 2　Metal Alloys

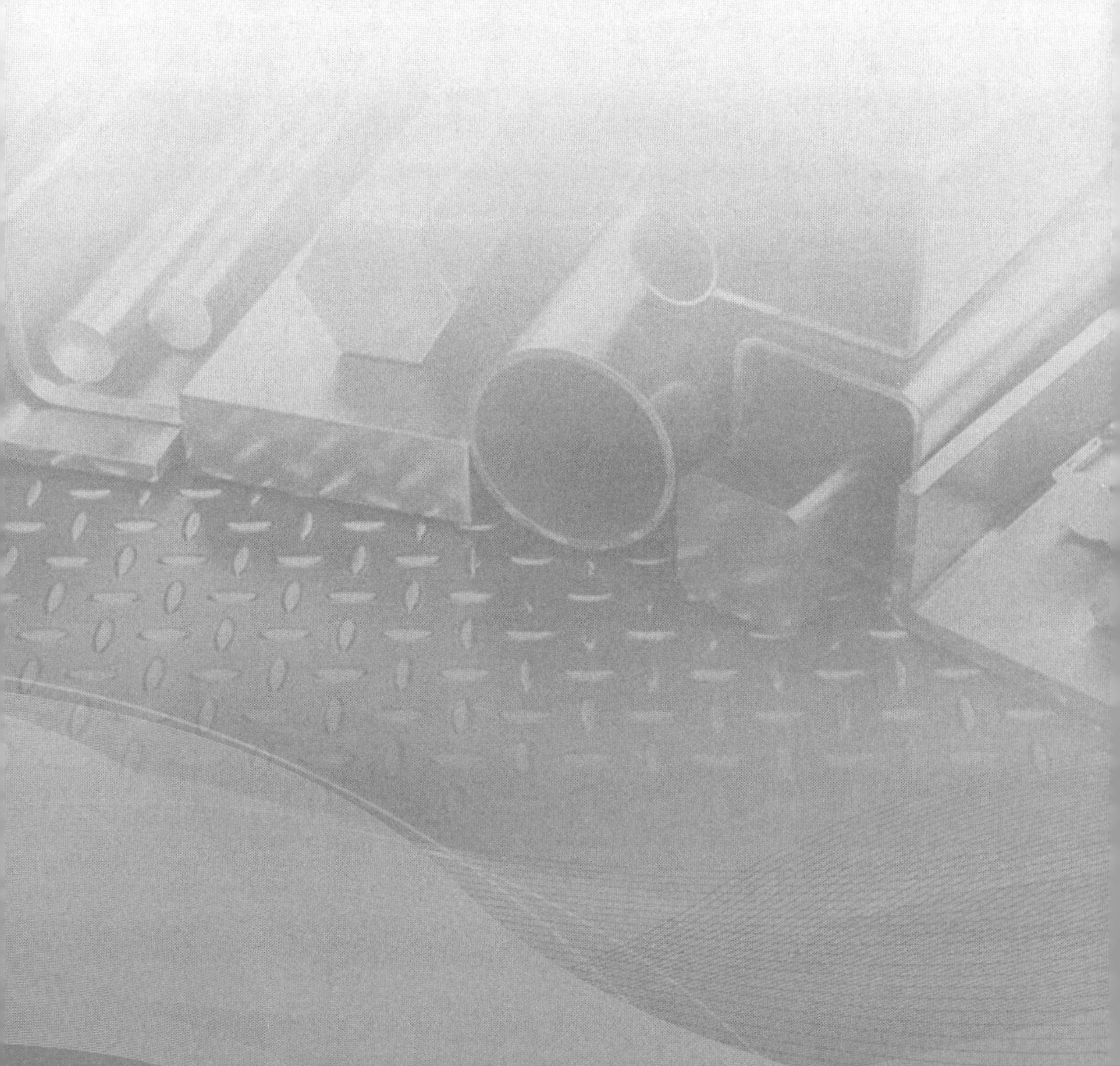

单元6 合 金
Unit 6 Alloys

扫码查看
讲课视频

6.1 教学内容
6.1 Teaching Materials

There are 90 or so naturally occurring elements and the majority of them are metals. But, useful though metals are, they're sometimes less than perfect for the jobs we need them to do. Take iron, for example. It's amazingly strong, but it can be quite brittle and it also rusts easily in damp air. Or what about aluminum. It's very light but, in its pure form, it's too soft and weak to be of much use. That's why most of the "metals" we use are not actually metals at all but alloys: metals combined with other substances to make them stronger, harder, lighter, or better in some other way. Alloys are everywhere around us—from the fillings in our teeth and the alloy wheels on our cars to the space satellites whizzing over our heads.

1. What is an Alloy?

You might see the word alloy described as a "mixture of metals", but that's a little bit misleading because some alloys contain only one metal and it's mixed in with other substances that are nonmetals (cast iron, for example, is an alloy made of just one metal, iron, mixed with one nonmetal, carbon). The best way to think of an alloy is as a material that's made up of at least two different chemical elements, one of which is a metal. The most important metallic component of an alloy (often representing 90 percent or more of the material) is called the main metal, the parent metal, or the base metal. The other components of an alloy (which are called alloying agents) can be either metals or nonmetals and they're present in much smaller quantities (sometimes less than 1 percent of the total). Although an alloy can sometimes be a compound, it's usually a solid solution.

2. The Structure of Alloys

If you look at a metal through a powerful electron microscope, you can see the atoms inside arranged in a regular structure called a crystalline lattice. Imagine a small cardboard box full of marbles and that's pretty much what you'd see. In an alloy, apart from the atoms of the main metal, there are also atoms of the alloying agents dotted throughout the structure.

(1) Substitution Alloys

If the atoms of the alloying agent replace atoms of the main metal, we get what's called a substitution alloy. An alloy like this will form only if the atoms of the base metal and those of the alloying agent are of roughly similar size. In most substitution alloys, the constituent elements are quite near one another in the periodic table. Brass, for example, is a substitution alloy based on copper in which atoms of zinc replace 10 ~ 35 percent of the atoms that would normally be in copper, as shown in Figure 1(a).

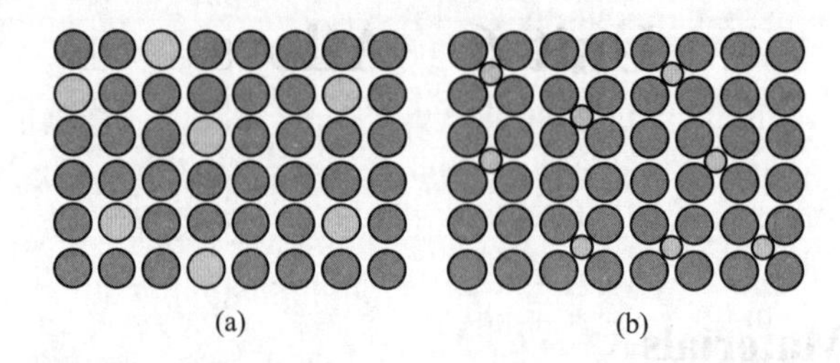

Figure 1 Substitution alloys(a) and interstitial alloys(b)
(The black circles represent the main metal and the red circles are the alloying agents)

(2) Interstitial Alloys

Alloys can also form if the alloying agent or agents have atoms that are very much smaller than those of the main metal. In that case, the agent atoms slip in between the main metal atoms (in the gaps or "interstices"), giving what's called an interstitial alloy. Steel is an example of an interstitial alloy in which a relatively small number of carbon atoms slip in the gaps between the huge atoms in a crystalline lattice of iron, as shown in Figure 1(b).

3. Properties of Alloys

People make and use alloys because metals don't have exactly the right properties for a particular job. Iron is a great building material but steel is stronger, harder, and rustproof. Aluminum is a very light metal but it's also very soft in its pure form. Add small amounts of the metals magnesium, manganese, and copper and you make a superb aluminum alloy called duralumin, which is strong enough to make airplanes. Alloys always show improvements over the main metal in one or more of their important physical properties (things like strength, durability, ability to conduct electricity, ability to withstand heat, and so on). Generally, alloys are stronger and harder than their main metals, less malleable (harder to work) and less ductile (harder to pull into wires).

4. The Fabrication of Alloys

The traditional way of making alloys was to heat and melt the components to make liquids, mix them together, and then allow them to cool into what's called a solid solution (the solid equivalent of a solution like salt in water). An alternative way of making an alloy is to turn the components into powders, mix them together, and then fuse them with a combination of high pressure and high temperature. This technique is called powder metallurgy. A third method of making alloys is to fire beams of ions into the surface layer of a piece of metal. Ion implantation, as this is known, is a very precise way of making an alloy. It's probably best known as a way of making the semiconductors used in electronic circuits and computer chips.

(Source: Chris Woodford *Alloys*. Explainthatstuff Website)

Words and Expressions

alloying *n.* 合金化处理；*a.*［材］合金的
rustproof *a.* 防锈的，不锈的
magnesium *n.*［化］镁（Mg）
duralumin *n.* 硬铝（一种铝合金）
component *n.* 组分，成分，部件
chip *n.* 芯片，碎屑，缺口
solid solution 固溶体
substitution alloy 置换合金，取代型合金
interstitial alloy 间隙合金
powder metallurgy 粉末冶金
main /parent /base metal 基体金属，母材

Phrases

made up of 由……组成
whizzing over 呼啸而过
full of 充满，装满
apart from 远离，除……之外
only if 仅当……

Notes

(1) Take iron, for example. It's amazingly strong, but it can be quite brittle and it also rusts easily in damp air. Or what about aluminum. It's very light but, in its pure form, it's too soft and weak to be of much use.

take…for example 译为"以…为例"；句子插入语较多，但前后两句结构一致，对比了铁铝的性能。

参考译文：以铁为例，它的强度惊人，但它也很脆，在潮湿的空气中也很容易生锈。铝呢？它很轻，但纯铝太软易变性，用处较少。

(2) You might see the word alloy described as a "mixture of metals", but that's a little bit misleading because some alloys contain only one metal and it's mixed in with other substances that are nonmetals (cast iron, for example, is an alloy made of just one metal, iron, mixed with one nonmetal, carbon).

you might see 后省略 that，引导了一个宾语从句；nonmetals 译为"非金属（元素）"；cast iron 译为"铸铁"。

参考译文：你可能会看到将合金描述为"混合金属"，但这有点误导，因为一些合金仅含一种金属，它是和另一种非金属元素混合后的产物（例如，铸铁是由金属铁与非金属碳制成的一种合金）。

(3) Steel is an example of an interstitial alloy in which a relatively small number of carbon atoms slip in the gaps between the huge atoms in a crystalline lattice of iron.

in which = where，引导定语从句，修饰 interstitial alloy（间隙型合金）。

参考译文：钢是一种间隙合金，在这种合金中，相对较少的碳原子在铁晶格形成的巨大原子间隙中移动。

(4) Alloys always show improvements over the main metal in one or more of their important physical properties (things like strength, durability, ability to conduct electricity, ability to withstand

heat, and so on).

over 译为“超过，优于”；physical properties 物理性能不止一种，故用复数；main metal 译为“基体金属，主体金属”。

参考译文：合金总是在一个或多个重要的物理性能（如强度、耐久性、导电能力、耐热能力等）上表现出比基体金属更好的性能。

(5) The traditional way of making alloys was to heat and melt the components to make liquids, mix them together, and then allow them to cool into what's called a solid solution (the solid equivalent of a solution like salt in water).

traditional 译为“传统的，习俗的，常用的”；allow to 译为“使之……，允许……”；solid solution 译为“固溶体”。

参考译文：传统制造合金的方法是将合金的成分加热熔化，制成液体，将它们混合在一起，然后让它们冷却成所谓的固溶体（一种相当于盐溶于水的固体溶液）。

6.2 阅读材料

6.2 Reading Materials

Mechanical Alloying

1. Introduction

The process of mechanical alloying (MA) starts with mixing of the powders in the desired proportion and loading of the powder mix into the mill along with the grinding medium (generally steel balls). Sometimes a process control agent (PCA) is added to prevent or minimize excessive cold welding of powder particles among themselves and/or to the milling container and the grinding medium. This mix (with or without the PCA) is then milled for the required length of time until a steady state is reached. At this stage alloying occurs and the composition of every powder particle is the same as the proportion of the elements in the starting powder mix. However, alloying is not required to occur during mechanical milling (MM), but only particle/grain refinement and/or some phase transformations should take place. The milled powder is then consolidated into a bulk shape and subsequently heat treated to obtain the desired microstructure and properties. Figure 1 shows a schematic of the different steps involved in preparing a component starting from the constituent powders by the process of MA. Thus, the important components of the MA process are the raw materials, the mill, and the process variables.

2. Raw Materials

The raw materials used for MA are the widely available commercially pure powders that have particle sizes in the range of 1 ~ 200μm. However, the powder particle size is not very critical, except that it should be smaller than the grinding ball size. This is because the powder particle size decreases exponentially with milling time and reaches a small value of a few micrometers only after a short period (typically a few minutes) of milling.

The raw powders fall into the broad categories of pure metals, master alloys, prealloyed powders,

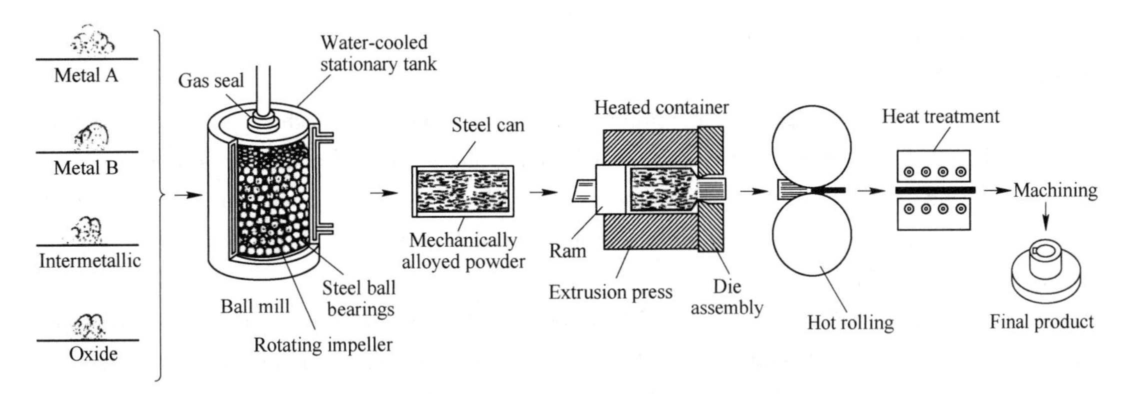

Figure 1 Schematic of the different steps involved in producing a product from powders by the mechanical alloying route

and refractory compounds. The oxygen content of the commercially pure metal powders ranges from 0.05 wt% to 2 wt%. Therefore, if one is interested in studying phase transformations in the milled powders, it is necessary to choose reasonably high-purity powders for the investigations. This is important because most commonly the nature and amount of impurities in the system decides the type of the final phase formed, chemical constitution of the alloy, and the kinetics of transformations.

Dispersion-strengthened materials usually contain additions of oxides, carbides, and nitrides. Oxides are the most common and these alloys are known as oxide dispersion strengthened (ODS) materials. In the early days of MA (the 1970s), the powder charge for MA consisted of at least 15 vol% of a ductile, compressibly deformable metal powder to act as a host or a binder. However, in recent years, mixtures of fully brittle materials have been milled successfully resulting in alloy formation. Thus, the earlier requirement of having a ductile metal powder during milling is no longer necessary. Consequently, ductile-ductile, ductile-brittle, and brittle-brittle powder mixtures have been milled to produce novel alloys. Recently, mixtures of solid powder particles and liquids have also been milled. In these cases, the liquid phase participates in alloying with the powder particles. For example, copper (solid) and mercury (liquid) have been milled together at room temperature to produce Cu-Hg solid solutions.

Occasionally, metal powder mixtures are milled with a liquid medium (here the liquid only facilitates milling but does not take part in alloying with the powder), and this is referred to as wet grinding; if no liquid is involved the process is called dry grinding. It has been reported that wet grinding is a more suitable method than dry grinding to obtain more finely ground products because the solvent molecules are adsorbed on the newly formed surfaces of the particles and lower their surface energy. The less agglomerated condition of the powder particles in the wet condition is also a useful factor. It has been reported that the rate of amorphization is faster during wet grinding than during dry grinding. However, a disadvantage of wet grinding is increased contamination of the milled powder. Thus, most of the MA/MM operations have been carried out dry. In addition, dry grinding is more efficient than wet grinding in some cases, e.g., during the decomposition of

$Cu(OH)_2$ to Cu under mechanical activation.

3. Types of Mills

Different types of high-energy milling equipment are used to produce mechanically alloyed/milled powders. They differ in their design, capacity, efficiency of milling, and additional arrangements for cooling, heating, and so forth. The following sections describe some of the more common mills currently in use for MA/MM, which are also readily available in the market.

(1) Spex Shaker Mills

Shaker mills, such as SPEX mills (Figure 2), which mill about 10~20g of the powder at a time, are most commonly used for laboratory investigations and for alloyscreen in purposes. The common version of the mill has one vial, containing the powder sample and grinding balls, secured in the clamp and swung energetically back and forth several thousand times a minute. The back-and-forth shaking motion is combined with lateral movements of the ends of the vial, so that the vial appears to be describing a figure of 8 or infinity symbol as it moves. Because of the amplitude (~5cm) and speed (~1200rpm) of the clamp motion, the ball velocities are high (on the order of 5m/s) and consequently the force of the ball's impact is unusually great. Therefore, these mills can be considered as high-energy variety.

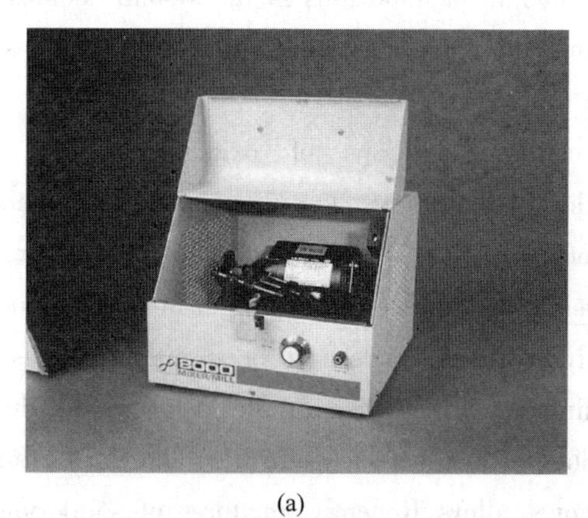

(a)

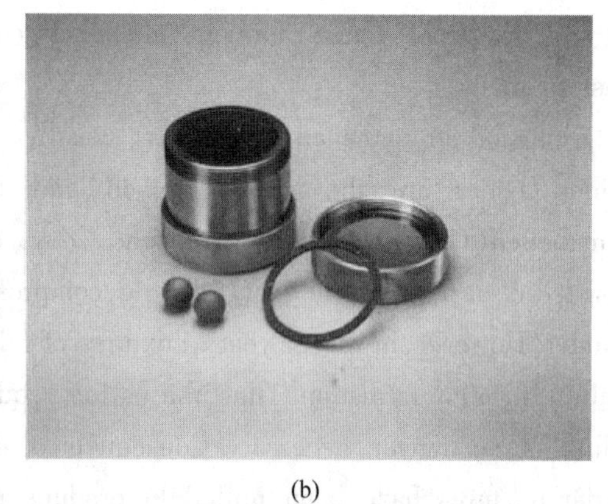

(b)

Figure 2　SPEX 8000 mixer/mill in the assembled condition (a), and tungsten carbide vial set consisting of the vial, lid, gasket and balls (b)

(2) Planetary Ball Mills

Another popular mill for conducting MA experiments is the planetary ball mill (referred to as Pulverisette) in which a few hundred grams of the powder can be milled at the same time (Figure 3). The planetary ball mill owes its name to the planet-like movement of its vials. These are arranged on a rotating support disk, and a special drive mechanism causes them to rotate around their own axes. The centrifugal force produced by the vials rotating around their own axes and that produced by the rotating support disk both act on the vial contents, consisting of the material to be ground and the grinding balls. The grinding balls in the planetary mills acquire much higher impact energy than is possible with simple pure gravity or centrifugal mills.

(3) Attritor Mills

An attritor (a ball mill capable of generating higher energies) consists of a vertical drum containing

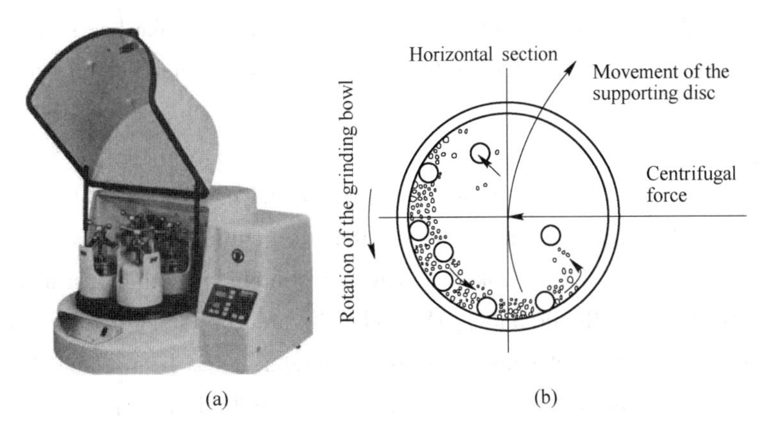

Figure 3 Fritsch Pulverisette P-5 four-station planetary ball mill(a), and Schematic depicting the ball motion inside the planetary ball mill(b)

a series of impellers(Figure 4). A powerful motor rotates the impellers, which in turn agitate the steel balls in the drum. Set progressively at right angles to each other, the impellers energize the ball charge. The dry particles are subjected to various forces such as impact, rotation, tumbling, and shear. This causes powder size reduction because of collisions between balls, between balls and container wall, and between balls, agitator shaft, and impellers. Therefore, micrometer-range fine powders can be easily produced.

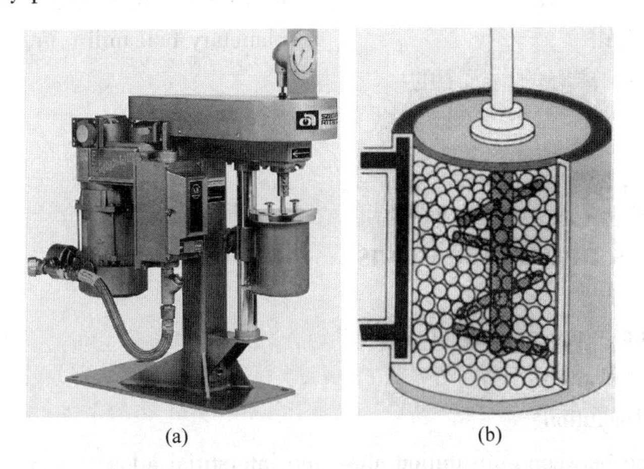

Figure 4 Model 01-HD attritor(a), and arrangement of rotating arms on a shaft in the attrition ball mill(b)

(Courtesy of Union Process, Akron, OH.)

(4) Commercial Mills

Commercial mills for MA are much larger in size than the mills described above and can process several hundred pounds of powder at a time. MA for commercial production is carried out in ball mills of up to about 3000lb(1250kg) capacity. The milling time decreases with an increase in the energy of the mill. It has been reported that 20min of milling in a SPEX mill is equivalent to 20h of milling in a low energy mill of the Invicta BX 920/2 type. As a rule of thumb, it can be estimated that a process that takes only a few minutes in the SPEX mill may take hours in an attritor and a few days in a commercial mill, even though the details can differ depending on the efficiency of the different mills.

(5) New Designs

Several new designs of mills have been developed in recent years for specialized purposes. These include the rod mills, vibrating frame mills, and the equipment available from Dymatron (Cincinnati, OH.), Nisshin Giken (Tokyo, Japan), Australian Scientific Instruments (Canberra, Australia), and M. B. N. srl (Rome, Italy). The rod mills are very similar to the ball mills except that they use steel rods instead of balls as grinding medium. It has been claimed that the powder contamination during rod milling is much lower than during ball milling. This has been explained to be due to the increased proportion of shear forces over the impact forces during rod milling.

(Source: Cury Suryanarayana. *Mechanical Alloying and Milling*. New York: Marcel Dekker, 2004)

Words and Expressions

schematic *a.* 概要的；*n.*（技术用语）示意图
variable *n.* ［数］变量；a. 多变的
grinding *n.* 研磨，磨削；磨碎
milling *n.* 铣削；*v.* 碾磨；滚（碾）轧
amorphization *n.* 无定形化
clamp *n.* 夹具，夹钳
centrifugal *a.* ［力］离心的，远中的
impeller *n.* ［机］叶轮，推进者
rotation *n.* 旋转
tumbling *n.* 翻滚，滚磨
collision *n.* ［物］碰撞，撞击
efficiency *n.* 效率，（机器的）功率
planetary ball mill 行星式球磨机

6.3 问题与讨论
6.3 Questions and Discussions

(1) Why not actually use pure metals?
(2) The definition of alloys.
(3) What is solid solution?
(4) The difference between substitution alloy and interstitial alloy.
(5) Introduce the fabrication methods of alloys.
(6) List some special properties of alloys.

单元7　金属合金制造
Unit 7　Production of Metal Alloys

扫码查看
讲课视频

7.1　教学内容
7.1　Teaching Materials

1. Alloying

There are many important reasons for alloying. The most common reason is to increase the mechanical properties of the base metal, such as higher strength, hardness, impact toughness, creep resistance and fatigue life. Alloying can also affect the other properties of metals: magnetic properties, electrical conductivity and corrosion resistance. In a few cases, alloying has the benefit of lowering the density of the metal, such as the addition of lithium to reduce the weight of aluminum alloys. Another reason for alloying is to increase the maximum working temperature of metals or improve their toughness at very low temperatures. Certain metals are alloyed to improve their corrosion resistance and durability in harsh environments. Alloying may also reduce the material cost when a cheap alloying element is added to an expensive base metal, although this should be considered a side benefit rather than the main reason for alloying.

In general, the effect of impurities is deleterious whereas the effect of alloying elements on properties is beneficial. Provided the concentration of impurity elements is kept low then they do not pose a problem and, in some materials, maybe beneficial to the processing or mechanical properties. Low concentrations of iron and atomic oxygen impurities in titanium increases the yield stress by solid solution strengthening. The process of adding alloy elements to a base metal is relatively simple. The base metal is melted inside a large crucible within a temperature-controlled furnace. The furnace environment is controlled during the melting of reactive base metals, such as titanium, nickel or magnesium, to stop oxidation and contamination from the air. Vacuum induction melting (VIM) is one of the most common methods. Alloying elements are added to the molten base metal in measured amounts to produce the metal alloy melt. Once the alloying elements have dissolved in the molten metal it is ready for casting.

2. Solubility of Alloying Elements

There are two types of solubility: unlimited solubility and limited solubility. Unlimited solubility means that the alloying element completely dissolves in the base metal, regardless of its concentration. For example, nickel has unlimited solubility in copper in concentrations less than 30% by weight. After solidification, the copper and nickel atoms do not separate but instead are dispersed throughout the material (Figure 1). The structure, properties and composition are uniform throughout the metal alloy. When this occurs the alloy is called a single-phase material.

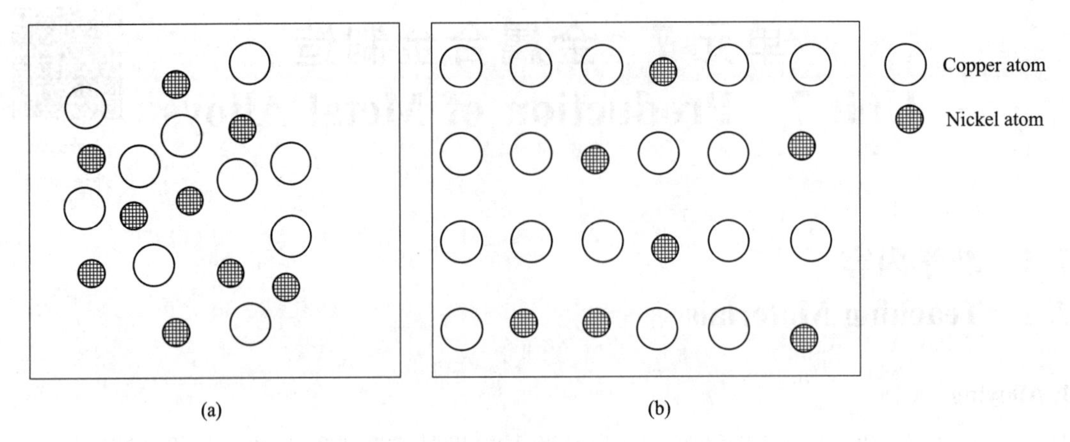

Figure 1 Unlimited solubility of nickel in copper
(a) Molten alloy; (b) Solid alloy

Unlimited solubility occurs when several conditions, known as the Hume-Rothery rules, are met.

(1) Size factor: The atomic sizes of the base metal and alloying element must be similar, with no more than a 15% difference to minimise the lattice strain.

(2) Crystal structure: The base metal and alloy element must have the same crystal structure.

(3) Electronegativity: The atoms of the base metal and alloying element must have approximately the same electronegativity.

Limited solubility means that an alloying element can dissolve into the base metal up to a concentration limit, but beyond this limit it forms another phase. The phase which is formed has a different composition, properties and crystal structure to the base metal. Most of the alloying elements used in aerospace base metals have low solubility. For example, when copper is present in aluminium it is soluble at room temperature up to about 0.2% by weight. At higher concentrations the excess of copper reacts with the aluminium to form another phase ($CuAl_2$). As another common example, when carbon is added to iron in the production of steel it has a solubility limit of under 0.005% by weight at 20℃. In higher concentrations, the excess of carbon atoms form carbide particles (e.g. Fe_3C) which are physically, mechanically and chemically different to the iron.

3. Selection of Alloying Elements

The types and amounts of alloying elements determine the metallurgical and mechanical properties of metals. Different alloying elements alter the properties in different ways. For example, when copper is added to aluminium it has a powerful strengthening effect but when used in titanium it has virtually no influence on the strength properties. As another example, chromium in steel promotes high corrosion resistance but has no impact on the corrosion properties of most other metals.

Certain alloying elements improve the strength properties by solid solution hardening or precipitation hardening, other elements increase the strength by refining the grain size, whereas different elements again may enhance the corrosion or oxidation resistance. For this reason, a number of alloying elements are used in the same metal rather than a single element. For instance, titanium alloys used in aircraft structures often contain two dominant alloying elements (aluminium and vanadium) with small concentrations of other elements (e. g. tin, zirconium, molybdenum). Some alloying elements have several functions, such as chromium in stainless steel that increases both strength and corrosion resistance.

The concentration of alloying elements is also critical to controlling the mechanical and durability properties. The properties of metals do not necessarily increase steadily with greater additions of alloying elements; instead there is an optimum concentration. For example, the optimum copper content in aluminium for maximum strengthening is 3% ~ 5%; below this range the metal is too soft and above this range too brittle.

(Source: Adrian P Mouritz. *Introduction to Aerospace Materials*. Cambridge: Woodhead Publishing Limited, 2012)

Words and Expressions

lithium *n.* ［化］锂（Li）
harsh *a.* 恶劣的，苛刻的
durability *n.* 耐久性
impurity *n.* 杂质，不纯，不洁
electronegativity *n.* 电负性
molten *a.* 熔化的，铸造的
nickel *n.* ［化］镍（Ni）
zirconium *n.* ［化］锆（Zr）
molybdenum *n.* ［化］钼（Mo）
carbide *n.* 碳化物
impact toughness 冲击韧性
creep resistance 抗蠕变
fatigue life 疲劳寿命
mechanical property 机械性能
grain size 晶粒大小
vacuum induction melting 真空感应熔炼
solid solution strengthening 固溶强化
precipitation hardening 沉淀硬化

Phrases

beneficial to 有益于……
regardless of 不管，不顾
be critical to 关键要素，对……起到关键作用
rather than 而不是，宁可……

Notes

(1) Alloying may also reduce the material cost when a cheap alloying element is added to an expensive base metal, although this should be considered a side benefit rather than the main reason for alloying.

base metal 译为“基底金属，母材”；side benefit 译为“额外好处，附加利益”；when 引导条件状语从句。

参考译文：当一种廉价的合金元素添加到一种昂贵的基体金属中时，合金化也可以降低材料成本，尽管这被认为是一种附带好处，而不是合金化的主要原因。

(2) Provided the concentration of impurity elements is kept low then they do not pose a

problem and, in some materials, maybe beneficial to the processing or mechanical properties.

provided 译为“假设”，为一条件状语；beneficial to 译为“有益于，有利于……”。

参考译文：只要将杂质元素的浓度保持在较低水平，那么它们就不会构成问题，并且在某些材料中可能对加工或机械性能有利。

(3) Alloying elements are added to the molten base metal in measured amounts to produce the metal alloy melt. Once the alloying elements have dissolved in the molten metal it is ready for casting.

molten 译为“熔融的”；once 引导条件状语；ready for 译为“准备好的，预备充分的”。

参考译文：在熔融的基体金属中加入一定量的合金元素以产生金属合金熔体。一旦合金元素溶解在熔化的金属中就可以准备铸造了。

(4) Limited solubility means that an alloying element can dissolve into the base metal up to a concentration limit, but beyond this limit it forms another phase. The phase which is formed has a different composition, properties and crystal structure to the base metal.

limit 在这里是名词“上限，极限”，而 limited 则是形容词“有限的”。

参考译文：有限溶解度是指合金元素在一定浓度范围内可以溶解到基体金属中，但超过这个范围就会形成另一种相。所形成的相具有与母材不同的成分、性质和晶体结构。

(5) Certain alloying elements improve the strength properties by solid solution hardening or precipitation hardening, other elements increase the strength by refining the grain size, whereas different elements again may enhance the corrosion or oxidation resistance.

solution hardening（固溶强化）、precipitation hardening（沉淀硬化）、refined strengthening（细晶强化）是金属强化的几种主要方式。

参考译文：某些合金元素通过固溶硬化或沉淀硬化来提高强度性能，其他元素通过细化晶粒尺寸来提高强度，而不同元素可以再次提高抗腐蚀或抗氧化性能。

7.2　阅读材料

7.2　Reading Materials

Solid Solutions

1. Solid Solutions

When homogeneous mixtures of two or more kinds of atoms occur in the solid state, they are known as solid solutions. These solutions are quite common and are equivalent to liquid and gaseous forms, for the proportions of the components can be varied within fixed limits, and the mixtures do not separate naturally. The term solvent refers to the more abundant atomic form, and solute to the less abundant. These solutions are also usually crystalline.

Solid solutions occur in either of two distinct types. The first is known as a substitutional solid solution. In this case, a direct substitution of one type of atom for another occurs so that solute atoms enter the crystal to take positions normally occupied by solvent atoms. Figure 1(a) shows schematically an example containing two kinds of atoms (Cu and Ni). The other type of solid solution is shown in Figure 1(b). Here the solute atom (carbon) does not displace a solvent atom, but, rather, enters one of the holes, or interstices, between the solvent (iron) atoms. This type of solution is known as an interstitial solid solution.

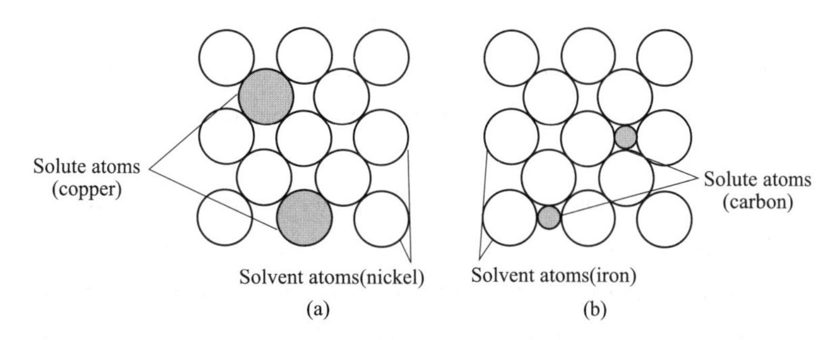

Figure 1 The two basic forms of solid solutions
(a) Substitutional solid solution; (b) Interstitial solid solution

2. Intermediate Phases

In many alloy systems, crystal structures or phases are found that are different from those of the elementary components (pure metals). If these structures occur over a range of compositions, they are, in all respects, solid solutions. However, when the new crystal structures occur with simple whole-number fixed ratios of the component atoms, they are intermetallic compounds with stoichiometric compositions.

The difference between intermediate solid solutions and compounds can be more easily understood by actual examples. When copper and zinc are alloyed to form brass, a number of new structures are formed in different composition ranges. Most of these occur in compositions which have no commercial value whatsoever, but that which occurs at a ratio of approximately one zinc atom to one of copper is found in some useful forms of brass. The crystal structure of this new phase is body-centered cubic, whereas that of copper is face-centered cubic, and zinc is close-packed hexagonal. Because this body-centered cubic structure can exist over a range of compositions (it is the only stable phase at room temperature between 47wt% and 50wt% of zinc), it is not a compound, but a solid solution. This is also sometimes called a nonstoichiometric compound or a nonstoichiometric intermetallic compound. On the other hand, when carbon is added to iron in an amount exceeding a small fraction of one-thousandth of a percent at ambient temperatures, a definite intermetallic compound is observed. This compound has a fixed composition (6.67 wt% of carbon) and a complex crystal structure (orthorhombic, with 12 iron atoms and 4 carbon atoms per unit cell) which is quite different from that of either iron (body-centered cubic) or carbon (graphite).

A binary alloy system may contain both stoichiometric and nonstoichiometric compounds. For example, aluminum and nickel alloys contain five nickel aluminide intermediate phases designated as Al_3Ni, Al_3Ni_2, AlNi and Al_3Ni_5, and $AlNi_3$. The first intermediate phase, Al_3Ni, has a fixed stoichiometric composition of 75 at.% Al and 25 at.% Ni. The other compounds, on the other hand, are nonstoichiometric.

3. Interstitial Solid Solutions

An examination of Figure 1 (b) shows that solute atoms in interstitial alloys must be small in size. The conditions that determine the solubilities in both interstitial and substitutional alloy systems have been studied in great detail by Hume-Rothery and others. According to their results, extensive interstitial solid solutions occur only if the solute atom has an apparent diameter smaller than 0.59 that of the solvent. The four most important interstitial solute atoms are carbon, nitrogen, oxygen, and hydrogen, all of which are small in size.

Atomic size is not the only factor that determines whether or not an interstitial solid solution will form. Small interstitial solute atoms dissolve much more readily in transition metals than in other metals. In fact, we find that carbon is so insoluble in most nontransitional metals that graphite-clay crucibles are frequently used for melting them. Some of the commercially important transition metals are Iron, Vanadium, Tungsten, Titanium, Chromium, Thorium, Zirconium, Manganese, Uranium, Nickel, Molybdenum.

The extent to which interstitial atoms can dissolve in the transition metals depends on the metal in question, but it is usually small. On the other hand, interstitial atoms can diffuse easily through the lattice of the solvent and their effects on the properties of the solvent are larger than might otherwise be expected. Diffusion, in this case, occurs not by a vacancy mechanism, but by the solute atoms jumping from one interstitial position to another.

4. Substitutional Solid Solutions and the Hume-Rothery Rules

In Figure 1(a), the copper and nickel atoms are drawn with the same diameters. Actually, the atoms in a crystal of pure copper have an apparent diameter (0.2551nm) about 2 wt% larger than those in a crystal of pure nickel (0.2487nm). This difference is small and only a slight distortion of the lattice occurs when a copper atom enters a nickel crystal, or vice versa, and it is not surprising that these two elements are able to crystallize simultaneously into a face-centered cubic lattice in all proportions. Nickel and copper form an excellent example of an alloy series of complete solubility.

Silver, like copper and nickel, crystallizes in the face-centered cubic structure. It is also chemically similar to copper. However, the solubility of copper in silver, or silver in copper, equals only a fraction of 1% at room temperature. There is, thus, a fundamental difference between the copper-nickel systems and the copper-silver systems. This dissimilarity is due primarily to a greater difference in the relative sizes of the atoms in the copper-silver alloys. The apparent diameter of silver is 0.2884nm, or about 13% larger than that of copper. It is thus very close to the limit noted first by Hume-Rothery, who pointed out that an extensive solid solubility of one metal in another only occurs if the diameters of the metals differ by less than 15%. This criterion for solubility is known as the size factor and is directly related to the strains produced in the lattice of the solvent by the solute atoms.

The size factor is only a necessary condition for a high degree of solubility. It is not a sufficient condition, since other requirements must be satisfied. One of the most important requirements is the relative positions of the elements in the electromotive series. (This series may normally be found in either a general chemistry text or an elementary book on the subject of corrosion.) Two elements, which lie far apart in this series, do not, as a rule, alloy in the normal sense, but combine according to the rules of chemical valence. In this case, the more electropositive element yields its valence electrons to the more electronegative element, with the result that a crystal with ionic bonding is formed. A typical example of this type of crystal is found in NaCl. On the other hand, when metals lie close to each other in the electromotive series, they tend to act as if they were chemically the same, which leads to metallic bonding instead of ionic.

Two other factors are of importance, especially when one considers a completely soluble system. Even if the size factor and electromotive series positions are favorable, such a system is only possible when both components (pure metals) have the same valence, and crystallize in the same lattice form.

(Source: Reza Abbaschian, et al. *Physical Metallurgy Principles*. Stamford: Gengage Learning, 2009)

Words and Expressions

stoichiometric *a.* 化学计量的，化学计算的
orthorhombic *a.* [晶] 正交晶的，斜方晶的
thorium *n.* [化] 钍 (Th)
uranium *n.* [化] 铀 (U)
electropositive *a.* 阳性的，带正电的
binary *a.* 二元的
electromotive series 元素电化序，电动序
substitutional solid solution 置换固溶体
interstitial solid solution 间隙固溶体
chemical valence 化学价
ionic bonding 离子键
metallic bonding 金属键
stable phase 稳定相

7.3 问题与讨论
7.3 Questions and Discussions

(1) What is the important reasons for alloying?
(2) Introduce one of the most common methods for alloying.
(3) The conditions for unlimited solubility.
(4) How to choose alloying elements.
(5) The principle of hardening by alloying.

单元 8　黑色金属
Unit 8　Ferrous Materials

扫码查看
讲课视频

8.1　教学内容
8.1　Teaching Materials

Ferrous materials containing iron, and steel is the most important material of this family. Because of availability, low cost, high strength, ease of fabrication into many shapes, and a wide range of properties, steel accounts for 80% of all metallic materials. It is widely used in manufacturing, building and construction industries. The demand of this material increases steadily with expansion of industries. In fact, the growth of steel production indicates the overall development of the industrial world. In 2013, total world crude steel production was 1607.2 million tons. China remains by far the largest producer of steel in the world.

The allotropic property of iron is responsible for obtaining different microstructures and properties. Below 910℃ (1670℉) iron is stable as α-phase (ferrite). Between 910 ~ 1392℃ (1670 ~ 2552℉), γ-phase (austenite) is stable, and above 1392℃ (2552℉) the δ-phase prevails. Iron is also an excellent solvent for many metals. The additions of the alloying elements to iron, especially carbon, stabilize these phases to a range of compositions and temperatures, which evolve the phase diagram for each individual system. The Fe-C system in Figure 1 shows how phase changes with variation of carbon content and temperature.

Pure iron is a soft material. By addition of carbon alone makes iron strong. Iron containing up to 2wt% carbon is called steel. When carbon content is over 2% carbon, the iron becomes very hard and can only be formed by casting to the required shape. This variety of iron is called cast iron. Steels have two broad groups: "plain carbon steels and alloy steels." Plain carbon steels contain C up to 1.7% with Si, Mn, S, P, Al, and some other elements below a critical limit where they do not affect the properties of the steel. The plain carbon steel is further divided as low, medium, and high carbon steels. Based on structure it is grouped as hypoeutectoid (ferrite + pearlite), eutectoid (pearlite), and hypereutectoid (pearlite + cementite) steels. However, the major shortcomings of plain carbon steels are low hardenability, loss of hardness on tempering, low corrosion and oxidation resistance, and low strength at elevated temperatures.

These limitations of plain carbon steel are overcome by alloying with other elements. By definition, the elements which are deliberately added to achieve desirable properties in a metallic system are called alloying elements, whereas those elements which are inevitably present in steel without any deliberate intention to serve additional purposes are known as impurities. For example,

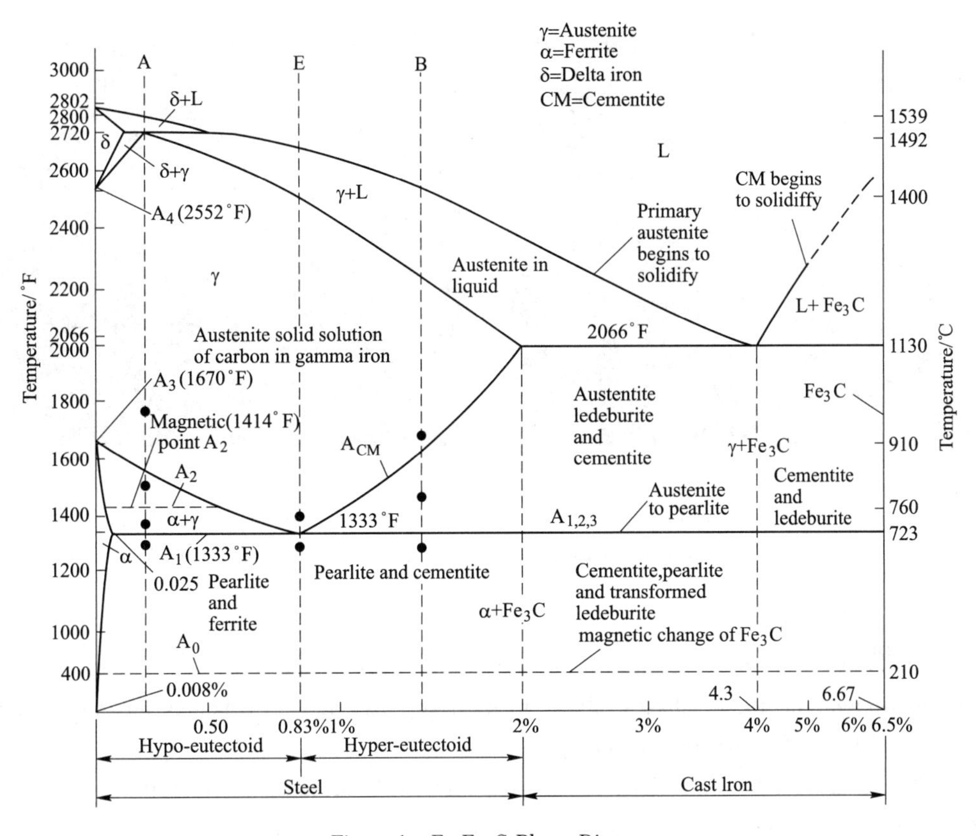

Figure 1 Fe-Fe_3C Phase Diagram

(Pollack, Prentice-Hall, *Materials secience and Metallurgy*, *4th ed.*, 1988)

the steel making operation cannot fully remove sulfur and therefore some amount of sulfur is always present in steel. Sulfur in such situation is called an impurity. Again the same sulfur when present in appreciable amount has the capacity to make chips brittle during machining. A particular variety of steel, known as free cutting steel contains deliberately added sulfur. In this case sulfur is an alloying element. Steels containing alloying element are called alloy steels. So long as the total alloying element content is less than 5 wt%, the steels are considered as low alloy steels whereas the steels with alloying elements in excess of 5 wt% is called high alloy steels. The alloy steels also identified on the basis of the added alloying element such as nickel steel, chromium steel, nickel-chromium steel, and so on. By structure it is known as martensitic steel, austenitic steel, ferritic steel, bainitic steel, and duplex steel. These groupings identify the range of properties and applications.

Development of microstructure involves some type of "phase transformation". Presence of these phases and morphology of these products decide the resultant properties of steel. Several heat treatment processes are employed to produce these phase structures in steels containing a range of compositions. The ferrite phase alone cannot be strengthened by heat treatment; cold working and grain size refining are the alternative options. Figure 2 shows such refined ferrite grain structure of hot rolled mild steel which can be achieved by changing the finishing temperature in hot rolling operation.

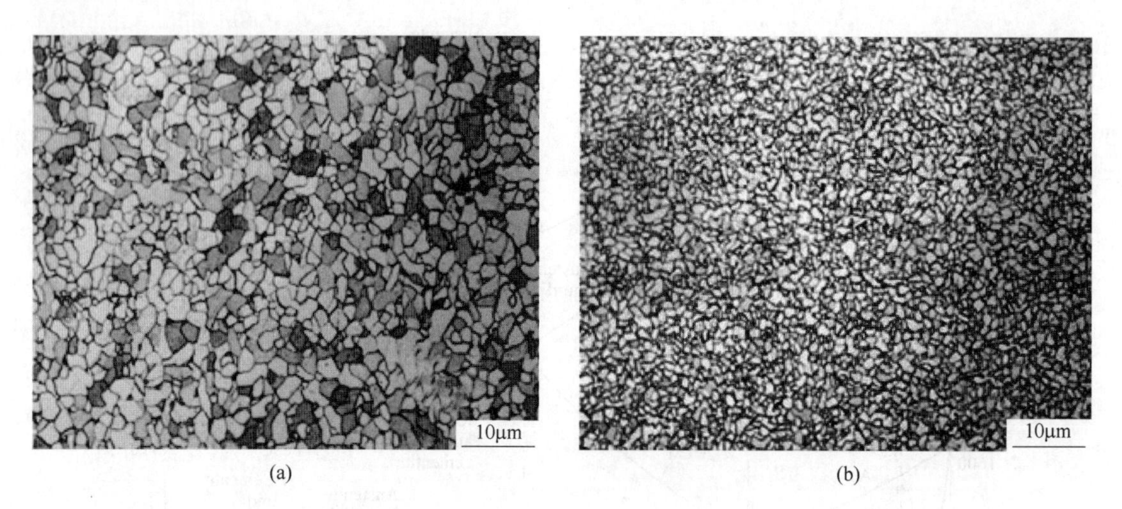

(a) (b)

Figure 2 Effect of finishing temperature on the ferrite grain structure of hot rolled mild steel
(a) Temperature above austenite transformation line, Al, (b) Temperature closer to Al line (Dasarathy, 2012)

Pearlite is an equilibrium structure and it is a combination of ferrite and cementite (Fe_3C) phases. Cementite is very hard, ferrite is soft and ductile while pearlite is very tough and strong. Distribution of these phases along with grain size and morphology of microstructure determine the level of properties in steels.

Cast irons, a distinct family of ferrous materials have wide diversity of properties. In the general grade of cast irons, carbon is present as combined carbon (Fe_3C) or free carbon (graphite). Combined carbon (cementite) makes the structure very hard, while free carbon, graphite gives excellent machining property. Ductile iron has a number of properties similar to those of steels yet it has advantages of low melting point, good fluidity and castability, excellent machinability, and good wear resistance.

(Source: S Mridha. *Matellic Materials*. Materials Science and Materials Engineering, 2016)

Words and Expressions

ferrous *a.* 铁的，含铁的
availability *n.* 可用性，有效性
allotropic *a.* 同素异形的
ferrite *n.* 铁素体
austenite *n.* 奥氏体
eutectoid *a.* 共析的
hypoeutectoid *a.* 亚共析的
hypereutectoid *a.* 过共析体的
martensitic *a.* 马氏体的
pearlite *n.* 珠光体
cementite *n.* 渗碳体，碳化铁
bainitic *a.* 贝氏体的
carburizing *n.* 渗碳
castability *n.* 铸造性能，［机］可铸性
graphite *n.* 石墨
phase diagram 相图
plain carbon steel 普通碳钢，碳素钢
duplex steel 二连钢，双炼钢，精炼钢
heat treatment 热处理
cold working 冷加工
hot rolling 热轧
cast iron 铸铁，生铁
critical limit 临界极限
ductile iron 球磨铸铁，韧性铁
wear resistance 耐磨性，耐磨度

Phrases

account for　对……负有责任，对……做出解释，导致
be responsible for　对……负责，是……的原因
have advantages of　具有……优点
similar to　和……相似

Notes

(1) Because of availability, low cost, high strength, ease of fabrication into many shapes, and a wide range of properties, steel accounts for 80% of all metallic materials.

account for 此处指“占……比例”。

参考译文：由于易得、廉价、高强，易于制造成多种形状以及广泛的性能，钢占所有金属材料的 80%。

(2) The additions of the alloying elements to iron, especially carbon, stabilize these phases to a range of compositions and temperatures, which evolve the phase diagram for each individual system.

addition to 译为“引入，加入”；which 引导定语从句；phase diagram 译为“相图”。

参考译文：铁中添加的合金元素，特别是碳，使这些相稳定在一定的组成和温度范围内，从而为每个个体系统演化成相图。

(3) By definition, the elements which are deliberately added to achieve desirable properties in a metallic system are called alloying elements, whereas those elements which are inevitably present in steel without any deliberate intention to serve additional purposes are known as impurities.

which 引导定语从句，修饰先行词 elements；whereas 译为“然而”，表示转折意义。

参考译文：根据定义，在金属体系中有意添加以获得理想性能的元素称为合金元素，而不可避免地存在于钢中的其他无特殊用途的元素称为杂质。

(4) Several heat treatment processes are employed to produce these phase structures in steels containing a range of compositions. The ferrite phase alone cannot be strengthened by heat treatment; cold working and grain size refining are the alternative options.

employed to 译为“被用于……”；heat treatment 译为“热处理”；cold working 译为“冷加工”。

参考译文：几种热处理工艺被用来在含有不同成分的一系列钢中产生这些相组织。铁素体相本身不能通过热处理得到强化；冷加工和晶粒细化是可供替代的方法。

(5) Ductile iron has a number of properties similar to those of steels yet it has advantages of low melting point, good fluidity and castability, excellent machinability, and good wear resistance.

ductile iron 专业术语“球墨铸铁”；have advantages of 译为“具有……优点”。

参考译文：球墨铸铁具有许多与钢相似的性能，还具有熔点低、流动性和铸造性好，优异的可加工性，以及好的耐磨性等优点。

8.2 阅读材料

8.2 Reading Materials

Iron-Carbon Phase Diagram

1. General Construction

The phase composition depending on the temperature and the carbon content can be read off this dual diagram as shown in Figure 1 in Section 8.1.

The closeness of the equilibrium lines which correspond to one another indicates that the difference in the stability of carbide and graphite in the alloys is not large. Therefore, the carbon may be dissolved in the iron after solidification or, however, may precipitate in the form of graphite. Furthermore, it can also occur in the structure in a bound form as iron carbide (Fe_3C) and is dissolved in the α and γ solid solution in both systems. The eutectic temperature of the iron-graphite reaction is considerably higher than that of the iron-cementite reaction. In the case of slower cooling, mainly graphite forms, in the case of accelerated heat dissipation, on the other hand, mainly cementite. During annealing, graphite may form, reducing the cementite content. The tendency for graphite to form from cementite shows that iron or iron-rich solid solutions only form a stable equilibrium with free carbon (graphite).

In the case of pure iron, more transformations take place during the heating and cooling in the solid state. It has become standard practice to label the which correspond to the transformations with A and to distinguish between them with numbers (A_1, A_2, A_3, A_4).

The types of crystal which are resistant in the different temperature ranges are labeled with Greek letters. On cooling curves, the halts are usually found at lower temperatures than on heating curves; the halts on the cooling curve are labeled as A_r points (r = refroidissement), those on the heating curve are labeled as A_c points (c = chauffage). The temperature difference between the respective halt temperatures during heating and cooling, the so-called hysteresis, is increased with faster temperature change. With very slow cooling, the positions of the halts become nearer according to the values given in the following as the equilibrium temperature:

(1) 1536℃ (2802°F): Solidification temperature (melting point), δ iron (Delta iron);

(2) 1392℃ (2552°F): A_4 point, γ iron (Gamma iron);

(3) 911℃ (1670°F): A_3 point, non-magnetic α iron;

(4) 769℃ (1414°F): A_2 point, ferromagnetic α iron;

(5) 723℃ (1333°F): A_1 point, pearlite point (eutectoid transformation, eutectoid temperature).

The cubic face-centered γ lattice can be described as a close sphere packing, the body-centered lattice of the δ and α iron is less closely packed (Delta iron). Accordingly, the volume of iron becomes smaller in the transformation $\alpha \rightarrow \gamma$, i. e. during heating at the A_3 point; however, it increases again dramatically in the transformation $\gamma \rightarrow \delta$, i. e. during heating at the A_4 point. The solvating power of iron for carbon depends on the type of lattice and the temperature. The face-centered γ iron has a must higher solvating power than the body-centered α iron.

2. The Iron Carbide(Iron Cementite) System

Iron carbide (cementite, Fe_3C) is unstable at all temperatures. Depending on the temperature, annealing can cause a disintegration of its austenite or ferrite, on the one hand, and graphite, on the other (austenite = γ solid solution; ferrite = α solid solution).

This is the basis for the production of blackheart malleable cast iron and the softening and ferritizing annealing. The cementite is extraordinarily resistant at room temperature. The eutectic temperature of the metastable system is 1147℃ and the eutectic composition is 4.26% C or 64% Fe_3C. This eutectic is also known as ledeburite; it consists of a set mixture of cementite and austenite (Figure 1).

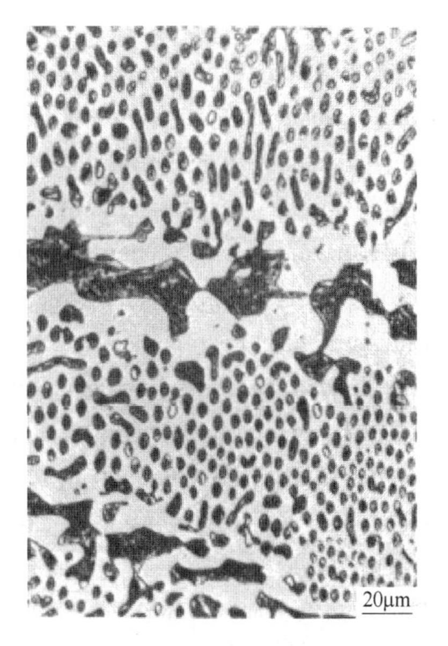

Figure 1 Ledeburite structure
(Black areas: cementite; white areas: ferrite; black patches in the middle: graphite, 500 : 1)

During further cooling, the austenite precipitates out of the secondary cementite which accumulates on the cementite which is already present and which can hardly be distinguished in the micrograph. The eutectoid transformation of the austenite into pearlite happens at 723℃. Consequently, the pearlite is made of α solid solution (ferrite) and cementite (Figure 2).

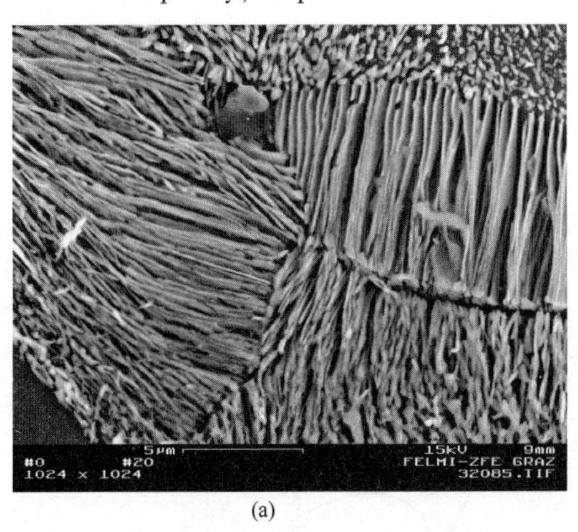

(a)

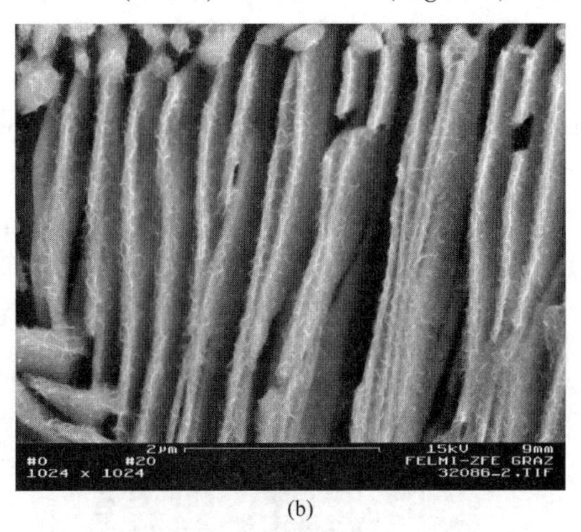

(b)

Figure 2 Pearlite formation in unalloyed GJS, wall thickness 50mm, etched (a), and detail from Figure 1 with EDX positions (b)

3. The Iron-Graphite System

All of the carbon is precipitated as graphite during solidification, whereby the eutectic is called graphite eutectic here. During further cooling, the secondary graphite precipitates out of the austenite which accumulates on the graphite lamellae and which cannot be distinguished. The austenite also undergoes eutectoid decomposition here, specifically into ferrite and graphite. In ferritic cast iron, the structure consists of a purely ferritic metal matrix with graphite embedded in it.

Generally, a gray solidification of cast ion can be forced according to the stable system. A eutectoid transformation into purely ferrite and graphite, on the other hand, is more difficult, as with progressive cooling, the inertia of the graphite crystallization increases considerably. Therefore, the eutectoid decomposition of the austenite rapidly changes in the meta-stable system, whereby pearlite is formed which is then reflected in a pearlitic metal matrix with flake graphite surrounded by ferrite borders. Pearlite-stabilizing additives in cast iron enable the achievement of a purely pearlitic structure (Ferritic cast iron, Ferrite border).

4. The Transformations during Cooling

The above described transformations of the austenite under equilibrium conditions only occur during slow cooling. By increasing the cooling rate, the transformations are deferred to lower temperatures. Due to under-cooling, they then often occur in temperature ranges in which the diffusion processes required to create the equilibrium can only happen incompletely or not at all. The transformation then occurs via meta-stable intermediate states, i. e. states which do not correspond to the equilibrium and which only have a more or less large stability at low temperatures. At the same time, very substantial changes occur in the process of the transformation and in the formation of the structure which forms.

With an increasing cooling rate, the melt is more and more under-cooled and when a certain critical cooling rate is exceeded, the stable system changes into the meta-stable system and chill occurs in cast iron (Figure 3). The eutectic solidification of a cast iron melt and the austenitic transformation represent a separation of a homogeneous phase into two phases: into austenite + graphite or cementite. The chill begins at the moment when, in a temperature range, the

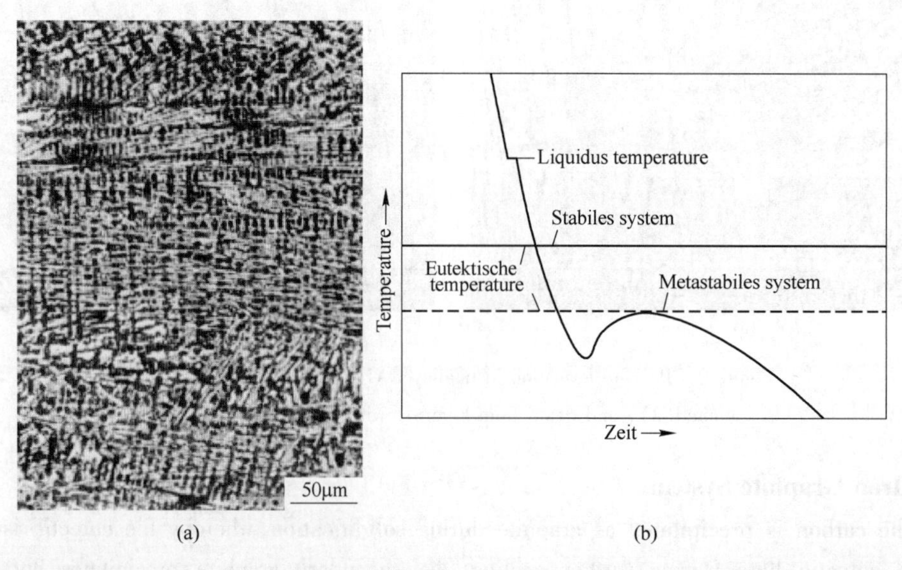

Figure 3　Structure of a chilled cast iron (a), and cooling curve of a completely chilled cast iron (b)

crystallization speed of the austenite-graphite eutectic is low than that of the austenite-cementite eutectic.

The chill can be prevented principally by the targeted addition of nuclei for the graphite, particularly by the addition of silicon (Inoculation). The presence of silicon means that the solidification favors the stable system (austenite/graphite) because the equilibrium temperature for the meta-stable system is reduced.

The gray solidification according to the stable system can be encouraged by slow cooling and by additives in the cast iron melt which have a graphitizing effect which increase the temperature range between the stable and meta-stable eutectic solidification and reduce the carbon solubility [silicon (Si), aluminum (Al), copper (Cu), nickel (Ni)]. Carbide-forming elements such as manganese (Mn), chromium (Cr), molybdenum (Mo), tungsten (W), tantalum (Ta), vanadium (V) and niobium (Nb), on the other hand, increase the solubility of the carbon and increase the tendency to chill, whereby, at the same time, the number of eutectic grains decreases (Eutectic grain, Eutectic grain count). In addition, however, there are also melt additives which have a graphitizing effect during solidification but have the opposite effect during austenitic transformation, i. e. pearlitizing, such as copper and nickel.

For the given casting wall thickness (constant cooling rate), by choosing a suitable composition and melt treatment, a chilled, mottled or gray cast iron with a ferritic or pearlitic metal matrix can be obtained (Metal matrix of cast iron). The solidification process of cast iron is also dependent on the number of nuclei in the melt and the growth rate of the eutectic grains (Balance of nuclei).

(Source: *Iron-Carbon phase diagram*. Foundry-Lexicon Website)

Words and Expression

solidification *n.* 固化，凝固

eutectic *a.* 共熔的，共晶的

graphitizing *n.* ［冶］石墨化作用

inoculation *n.* ［医］接种，接木，接插芽

cooling rate 冷却速度

8.3 问题与讨论

8.3 Questions and Discussions

(1) The allotropic properties of iron?

(2) Master the Fe-C phase diagram.

(3) What is steel and what is cast iron?

(4) The classification of plain carbon steel.

(5) How to strengthen ferrite phase?

(6) What properties do ductile iron have?

单元9 有色金属
Unit 9 Nonferrous Metals

扫码查看
讲课视频

9.1 教学内容
9.1 Teaching Materials

When a metal is defined as non-ferrous it means that it does not have a significant amount of iron in its chemical composition. Non-ferrous alloys generally have iron compositions of less than one percent as measured by weight. In metallurgy, a non-ferrous metal is a metal, including alloys, that does not contain iron (ferrite) in appreciable amounts. Generally more costly than ferrous metals, non-ferrous metals are used because of desirable properties such as low weight (e. g. aluminium), higher conductivity (e. g. copper), non-magnetic property or resistance to corrosion (e. g. zinc). Some non-ferrous materials are also used in the iron and steel industries. For example, bauxite is used as flux for blast furnaces, while others such as wolframite, pyrolusite and chromite are used in making ferrous alloys.

Important non-ferrous metals include aluminium, copper, lead, nickel, tin, titanium and zinc, and alloys such as brass. Precious metals such as gold, silver and platinum and exotic or rare metals such as cobalt, mercury, tungsten, beryllium, bismuth, cerium, cadmium, niobium, indium, gallium, germanium, lithium, selenium, tantalum, tellurium, vanadium, and zirconium are also non-ferrous. They are usually obtained through minerals such as sulfides, carbonates, and silicates. Non-ferrous metals are usually refined through electrolysis.

1. Ancient History

Non-ferrous metals were the first metals used by humans for metallurgy. Gold, silver and copper existed in their native crystalline yet metallic form. These metals, though rare, could be found in quantities sufficient to attract the attention of humans. Less susceptible to oxygen than most other metals, they can be found even in weathered outcroppings. Copper was the first metal to be forged; it was soft enough to be fashioned into various objects by cold forging and could be melted in a crucible. Gold, silver and copper replaced some of the functions of other resources, such as wood and stone, owing to their ability to be shaped into various forms for different uses. Due to their rarity, these gold, silver and copper artifacts were treated as luxury items and handled with great care. The use of copper also heralded the transition from the Stone Age to the Copper Age. The Bronze Age, which succeeded the Copper Age, was again heralded by the invention of bronze, an alloy of copper with the non-ferrous metal tin.

2. Common Properties and Structural Uses

It is nearly impossible to define the common properties of non-ferrous metals simply because there is such a large variety of metals that fall into the non-ferrous category. Some non-ferrous metals are hard and brittle, some soft and ductile. Some non-ferrous metals are made for cryogenic applications, others are made to withstand extremely high temperatures. There are far more differences than there are similarities among the different types of non-ferrous metals. However, non-ferrous metals all do have one thing in common: They do not rust. That is not to say that they don't corrode. Another characteristic of non-ferrous metals is that they are not magnetic.

It is used in residential, commercial, industrial industry. Material selection for a mechanical or structural application requires some important considerations, including how easily the material can be shaped into a finished part and how its properties can be either intentionally or inadvertently altered in the process. Depending on the end use, metals can be simply cast into the finished part, or cast into an intermediate form, such as an ingot, then worked, or wrought, by rolling, forging, extruding, or other deformation process.

3. Important Non-Ferrous Metals

(1) Aluminum

Aluminum is a very widely used type of non-ferrous alloy. In its unanodized form, it has a silvery color. Without the addition of alloying elements, it is more ductile and not quite as strong as many steels. However, through the addition of alloying elements and heat treating or work hardening, aluminum can achieve very high strengths. Aluminum is lighter than steel. It forms a protective oxide layer that helps it reduce the risk of detrimental corrosion. Common applications of aluminum include marine equipment such as boat lifts and docks; aerospace equipment such as airplane body material; construction material such as beams and rails; and certain types of cookware.

(2) Copper

Copper is another very popular non-ferrous alloy. In its unalloyed state, it too is softer, more ductile, and not as strong as carbon steel. However, similar to aluminum, copper can be alloyed with a variety of elements to give it improved mechanical properties. When copper is alloyed with tin it is considered bronze. When copper is alloyed with significant amounts of zinc, it is called brass. Common applications of pure copper and its alloyed forms include electrical components such as wires, terminals, and other types of connectors; currency such as the United States or Canadian penny (although just as a coating); pipe for plumbing, tooling, and decorative work.

(3) Nickel

Nickel is another popular non-ferrous alloy. Nickel is known for its toughness, ability to perform in high temperature and low temperature environments, and corrosion resistance. Nickel is not often used in its pure form, and like copper and aluminum, it is often alloyed with other elements to gain superior chemical and mechanical properties. Common applications of nickel and nickel-based alloys include cryogenic equipment such as tanks; hot-section aerospace equipment such as combustion chamber components; and marine equipment.

This is not an exhaustive list of non-ferrous metals. There are many more such as tungsten, titanium, zinc, silver, gold, platinum and lead.

(Source: *Non-ferrous Metal.* Metalsupermarkets Website)

Words and Expressions

constitute　*v.* 构成，组成
wolframite　*n.* 黑钨矿，钨锰铁矿
pyrolusite　*n.* 软锰矿
chromite　*n.* 铬铁矿，亚铬酸盐
cobalt　*n.* ［化］钴（Co）
mercury　*n.* 水银，［化］汞（Hg）
tungsten　*n.* ［化］钨（W）
beryllium　*n.* ［化］铍（Be）
bismuth　*n.* ［化］铋（Bi）
cadmium　*n.* ［化］镉（Cd）
callium　*n.* ［化］镓（Ga）
germanium　*n.* ［化］锗（Ge）
selenium　*n.* ［化］硒（Se）
tantalum　*n.* ［化］钽（Ta）
tellurium　*n.* ［化］碲（Te）
cerium　*n.* ［化］铈（Ce）
niobium　*n.* ［化］铌（Nb）
indium　*n.* ［化］铟（In）
sulfide　*n.* 硫化物
carbonate　*n.* 碳酸盐
silicate　*n.* 硅酸盐
electrolysis　*n.* 电解，电解作用
forge　*v.* 锻造（金属）
crucible　*n.* 坩埚
cryogenic　*a.* 低温的，冷冻的
inadvertently　*ad.* 无意地，不经意地
ingot　*n.* 锭，铸块
rolling　*n.* 滚动，旋转
forging　*n.* 锻造，锻件
marine　*a.* 海洋的，海运的，海事的
chemical composition　化学组成
rare metal　稀有金属

Phrases

far more difference　　更差
be defined as　　定义为……

Notes

(1) Generally more costly than ferrous metals, non-ferrous metals are used because of desirable properties such as low weight (e. g. aluminium), higher conductivity (e. g. copper), non-magnetic property or resistance to corrosion (e. g. zinc).

desirable 译为“令人向往的，值得拥有的，好的，理想的，想要的”；non-ferrous metals 译为“有色金属”，与 ferrous motal（黑色金属）做对比。

参考译文：一般来说，有色金属比黑色金属更昂贵，使用有色金属是由于其理想的特性，如重量轻（如铝），较高的导电性（如铜），无磁性或耐腐蚀（如锌）。

(2) It is nearly impossible to define the common properties of non-ferrous metals simply because there is such a large variety of metals that fall into the non-ferrous category.

fall into 译为“落入，分成，属于”；because 引导原因状语从句，后面 that 又引导一个宾语从句。

参考译文：因为有这么多的金属可归入有色金属类别，所以简单地定义有色金属的共性是不可能的。

(3) Material selection for a mechanical or structural application requires some important considerations, including how easily the material can be shaped into a finished part and how its

properties can be either intentionally or inadvertently altered in the process.

intentionally or inadvertently 译为“有意或无意”；finished 译为“已完成的，最终的”；shaped into 译为“成形”。

参考译文：机械或结构应用材料的选择需要考虑一些重要因素，包括材料如何容易成型，以及其性能如何在成形过程中被有意或无意地改变。

(4) In its unalloyed state, it too is softer, more ductile, and not as strong as carbon steel. However, similar to aluminum, copper can be alloyed with a variety of elements to give it improved mechanical properties.

unalloyed 译为“纯的，未被合金化的”，相当于“pure”。

参考译文：在它的纯合金状态下，它也更软，更有延展性，不像碳钢那样强。然而，与铝类似，铜可以与多种元素形成合金，从而提高其力学性能。

9.2　阅读材料
9.2　Reading Materials

Aluminum and Aluminum Alloys

1. Overview

Aluminum and aluminum alloys have many outstanding attributes that lead to a wide range of applications, including good corrosion and oxidation resistance, high electrical and thermal conductivities, low density, high reflectivity, high ductility and reasonably high strength, and relatively low cost.

Aluminum is a lightweight material with a density of 2.7g/cm^3(0.1lb/in^3). Pure aluminum and its alloys have the face-centered cubic(fcc) structure, which is stable up to its melting point at 657℃. Because the fcc structure contains multiple slip planes, this crystalline structure greatly contributes to the excellent formability of aluminum alloys. Aluminum alloys display a good combination of strength and ductility. Aluminum alloys are among the easiest of all metals to form and machine. The precipitation hardening alloys can be formed in a relatively soft state and then heat treated to much higher strength levels after forming operations are complete. In addition, aluminum and its alloys are nontoxic and among the easiest to recycle of any of the structural materials.

Aluminum is the most abundant metal in the Earth's crust, but it was not until the 1800s that elemental aluminum was success-fully extracted. Even the first processes developed were inefficient and extremely expensive. The situation changed in 1886 to 1888 with the nearly simultaneous development of the Hall-Héroult process for electrolytic reduction and the Bayer process for inexpensive production of alumina (Al_2O_3) from bauxite ore. These breakthroughs allowed the widespread production and use of aluminum and aluminum alloys. Charles Hall, the developer of the Hall-Héroult process, went on to form the Aluminum Company of America(Alcoa).

2. Types of Aluminum Alloys

Aluminum alloys are normally classified into one of three groups: wrought non-heat-treatable alloys, wrought heat treatable alloys, and casting alloys.

(1) Wrought non-heat-treatable alloys cannot be strengthened by precipitation hardening; they are hardened primarily by cold working. The wrought non-heat-treatable alloys include the commercially pure aluminum series (1×××), the aluminum-manganese series (3×××), the aluminum-silicon series(4×××), and the aluminum-magnesium series(5×××). While some of the 4xxx alloys can be hardened by heat treatment, others can only be hardened by cold working.

(2) Wrought heat treatable alloys can be precipitation hardened to develop quite high strength levels. These alloys include the 2 ××× series(Al-Cu and Al-Cu-Mg), the 6××× series(Al-Mg-Si), the 7××× series(Al-Zn-Mg and Al-Zn-Mg-Cu), and the aluminum-lithium alloys of the 8××× alloy series. The 2 ××× and 7××× alloys, which develop the highest strength levels, are the main alloys used for metallic aircraft structure.

(3) Casting alloys include both non-heat-treatable and heat treat-able alloys. The major categories include the 2××. × series(Al-Cu), the 3××. × series(Al-Si + Cu or Mg), the 4××. × series(Al-Si), the 5××. × series(Al-Mg), the 7××. × series(Al-Zn), and the 8××. × series(Al-Sn). The 2××. ×, 3××. ×, 7××. ×, and 8××. × alloys can be strengthened by precipitation hardening, but the properties obtained are not as high as for the wrought heat treatable alloys.

3. Properties

Among the most striking characteristics of aluminum is its versatility. More than 300 alloy compositions are commonly recognized, and many additional variations have been developed internationally and in supplier/consumer relationships. The properties of aluminum that make this metal and its alloys the most economical and attractive for a wide variety of uses are appearance, light weight, fabricability, physical properties, mechanical properties, and corrosion resistance.

Aluminum can display excellent corrosion resistance in most environments, including atmosphere, water (including salt water), petrochemicals, and many chemical systems. Aluminum surfaces can be highly reflective. Radiant energy, visible light, radiant heat, and electromagnetic waves are efficiently reflected, while anodized and dark anodized surfaces can be reflective or absorbent. The reflectance of polished aluminum, over a broad range of wavelengths, leads to its selection for a variety of decorative and functional uses. Aluminum is nonferromagnetic, a property of importance in the electrical and electronics industries. It is nonpyrophoric, which is important in applications involving inflammable or explosive materials handling or exposure. Aluminum is also nontoxic and is routinely used in containers for foods and beverages. It has an attractive appearance in its natural finish, which can be soft and lustrous or bright and shiny. It can be virtually any color or texture.

Aluminum typically displays excellent electrical and thermal conductivity, but specific alloys have been developed with high degrees of electrical resistivity. Aluminum is often selected for its electrical conductivity, which is nearly twice that of copper on an equivalent weight basis. The requirements of high conductivity and mechanical strength can be met by use of long-line, high-voltage, aluminum steel-cored rein-forced transmission cable. The thermal conductivity of aluminum alloys, approximately 50% to 60% that of copper, is advantageous in heat exchangers, evaporators, electrically heated appliances and utensils, and automotive cylinder heads and radiators.

Some aluminum alloys exceed structural steel in strength. However, pure aluminum and certain aluminum alloys are noted for extremely low strength and hardness. The tensile yield strength of superpurity aluminum in its softest annealed state is approximately 10MPa(1.5ksi), whereas that of some heat treated commercial high-strength alloys exceeds 550MPa(80ksi). Higher strengths, up to a yield strength of 690MPa(100ksi) and over, may be readily produced, but the fracture toughness of such alloys does not meet levels considered essential for aircraft or other critical-structure applications. The density of aluminum and its alloys is approximately 7GPa(10MSI), which is lower than titanium(10GPa, or 15MSI) and steel(21GPa, or 30MSI) alloys. However, when measured on a stiffness-to-density basis, aluminum alloys are weight-competitive with the heavier titanium and steel alloys. One rather disappointing property of high-strength aluminum alloys is their fatigue performance; the fatigue limit of most high-strength alloys falls within the 137 to 172MPa (20 to 25ksi) range. Work hardening raises the strength of aluminum quite substantially. Commercial-purity aluminum(99.60% pure) has a yield strength of 27MPa(3.9ksi) when fully annealed, but if cold worked by swaging or rolling to 75% reduction in area, the yield strength increases to 125MPa(18ksi).

4. Applications

Aluminum is a consumer metal of great importance. Aluminum and its alloys are used for foil, beverage cans, cooking and food processing utensils, architectural and electrical applications, and structures for boats, aircraft, and other transportation vehicles. Alloy 3004, which is used for beverage cans, has the highest single usage of any aluminum alloy, accounting for approximately 1/4 of the total usage of aluminum. Its corrosion and oxidation resistance is especially important in architectural and transportation applications. With a yield strength comparable to that of mild steel, 6061 is one of the most widely used of all aluminum alloys for general construction.

The 5××× alloys are used extensively in the transportation industries for boat and ship hulls; dump truck bodies; large tanks for carrying gasoline, milk, and grain; and pressure vessels, especially where cryogenic storage is required. The weldability of these alloys is excellent, and they have excellent corrosion resistance. Its high thermal conductivity leads to applications such as radiators and cooking utensils. Its low density is important for hand tools and all forms of transportation, especially aircraft.

The high-strength 2××× and 7××× alloys are competitive on a strength-to-weight ratio with the higher-strength but heavier titanium and steel alloys and thus have traditionally been the dominant structural material in both commercial and military aircraft. In addition, aluminum alloys are not embrittled at low temperatures and become even stronger as the temperature is decreased without significant ductility losses, making them ideal for cryogenic fuel tanks for rockets and launch vehicles. Aluminum-lithium alloys are attractive for aerospace applications because the addition of lithium increases the modulus of aluminum and reduces the density(each 1 wt% of lithium increases the modulus by approximately 6% while decreasing the density approximately 3%).

(Source: *Aluminum and Aluminum Alloys*. ASM International Website)

Words and Expressions

reflectivity *n.* 反射率
breakthrough *n.* 突破
wavelength *n.* 波长
resistivity *n.* [电]电阻率，抵抗力
nonpyrophoric *a.* 不发生火花的
texture *n.* 质地，纹理，口感
embrittle *v.* 使变脆，变脆
versatility *n.* 多功能性，用途广泛
fabricability *n.* 加工性，可制性
petrochemical *a.* 石油化工的
swaging *n.* 模锻，挤锻，型锻
radiator *n.* 暖气片，散热器，辐射源
foil *n.* 箔，箔纸
evaporator *n.* 蒸发器，蒸发设备
inflammable *a.* 易燃的 *n.* 易燃物
oxidation resistance 抗氧化性
structural material 结构材料
mechanical strength 机械强度

9.3 问题与讨论

9.3 Questions and Discussions

(1) List some important nonferrous metals.
(2) Common properties of nonferrous metals.
(3) How to process an ingot?
(4) What is the common applications of aluminum alloys?
(5) What is nickel known for?

第三部分　机械加工

Part 3　Mechanical Machining

单元 10　机械加工工艺
Unit 10　Machining Process

扫码查看
讲课视频

10.1　教学内容
10.1　Teaching Materials

The term machining deals with any process in which material is removed gradually from a workpiece, including metal cutting with both single-point or multi-point tools (tools with geometrically defined cutting edges) and grinding with abrasive wheels which consists of a large number of micro-cutting edges, randomly shaped and oriented, i. e. tools with geometrically undefined cutting edges. In the engineering industry, the term machining is used to cover chip-forming operations resulting in mass reducing. The narrower term cutting is intended to include operations in which a thin layer of material, the chip or swarf, is removed by a wedge-shaped tool(case Ⅰ in Table 1).

Table 1　Classification of machining processes based on the energy used to material removal proposed by Alting

Category of basic process		Fundamental removal method	Examples of processes
Mechanical	Ⅰ		Cutting Turning Milling Drilling Grinding, etc.
	Ⅱ		Water jet cutting Abrasive jet machining Sand blasting, etc.
	Ⅲ		Ultrasonic machining
	Ⅳ		Blanking Punching Shearing

Continued Table 1

Category of basic process		Fundamental removal method	Examples of processes
Thermal	Ⅱ		Thermal cutting(melting) Electron beam machining Laser machining
	Ⅲ		Electrodischarge machining(EDM)
Chemical	Ⅱ		Etching Thermal cutting(combustion)
	Ⅲ		Electrochemical machining(ECM)

Table 1 shows the classification of mass-reducing processes in terms of the process category and the method of material removal. Group Ⅰ in the mechanical category covers many popular conventional cutting processes, known also as machining processes with geometrically defined cutting edges. Other processes, which utilize mechanical, thermal and chemical energy (groups Ⅱ and Ⅲ which utilize highly concentrated energy beams or metal removal by dissolving or evaporating), are classified as non-conventional machining processes. There is no sharp line separating chip-forming operations from others such as shearing, blanking and punching (case Ⅳ) which are commonly classified as metal-forming operations. While metal cutting is commonly associated with big industries (automotive, aircraft, aerospace, home appliance, etc.) that manufacture big products, the machining of metals and alloys plays a crucial role in a range of manufacturing activities, including the ultraprecision machining of extremely delicate components.

The most commonly used criteria for selection of machining operations are the kinematical features and the capability of shaping appropriate surfaces, e. g. rotational, prismatic (planes), helical, threads, 3D flee-form (sculptured) surfaces. In this content we distinguish three basic machining operations, i. e. turning operations (Figure 1), drilling and related hole-making operations (Figure 2) and milling operations (Figure 3).

As shown in Figure 1, turning with single-point tools enables external and internal cylindrical surfaces (longitudinal turning) and taper surfaces, as well as 2D complex shapes using NC lathes equipped with contouring control systems, to be produced. It is also possible to perform facing, parting and cutoff and threading operations.

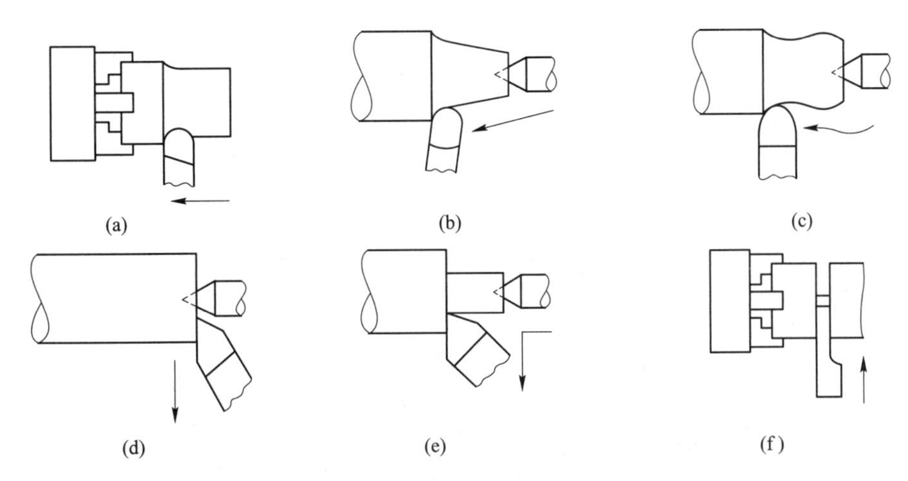

Figure 1　Some typical turning operations

(a) Straight turning; (b) Taper turning; (c) Contour turning; (d) End facing; (e) Shoulder facing; (f) Parting or cut off

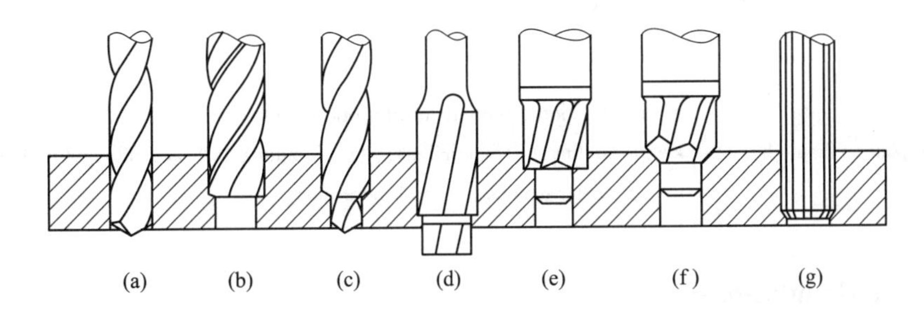

Figure 2　Common drilling and related operations

(a) Drilling; (b) Core drilling; (c) Step drilling; (d) Boring; (e) Counterboring; (f) Countersinking; (g) Reaming

Hole-making operations based on drilling and allied operations are presented in Figure 2. Drilling is the easiest way to make a hole in a solid metal with tolerances between IT13 and IT9(8). It is also applied to enlarge holes—a process called core drilling or counterdrilling. The operation of making the hole of two or more different diameters is called step drilling. Boring is the enlarging of a hole, typically resulting in obtaining higher accuracy and better surface finish than by drilling. Enlarging a hole to a limited depth is called counterboring. Producing an angular opening into the end of a hole is countersinking, also termed chamfering.

A range of milling operations shown in Figure 3 can be used to generate flat surfaces as well as 2D or 3D curved surfaces, but by using rotating cutters with multiple cutting edges or teeth. As indicated in Figure 3, the same kind of surface can be milled in several ways. For instance, plane surfaces may be produced by slab milling, face or side milling. Peripheral (plain or slab) milling generates a surface parallel to the axis of rotation while face milling generates a flat surface normal to the axis of rotation. End milling, a type of face milling operation, is used for facing, profiling and slotting operations, i. e. to produce one-, two- or three-sided surfaces. Now, the use of inserted teeth milling cutters has minimized the problem of milling machine maintenance.

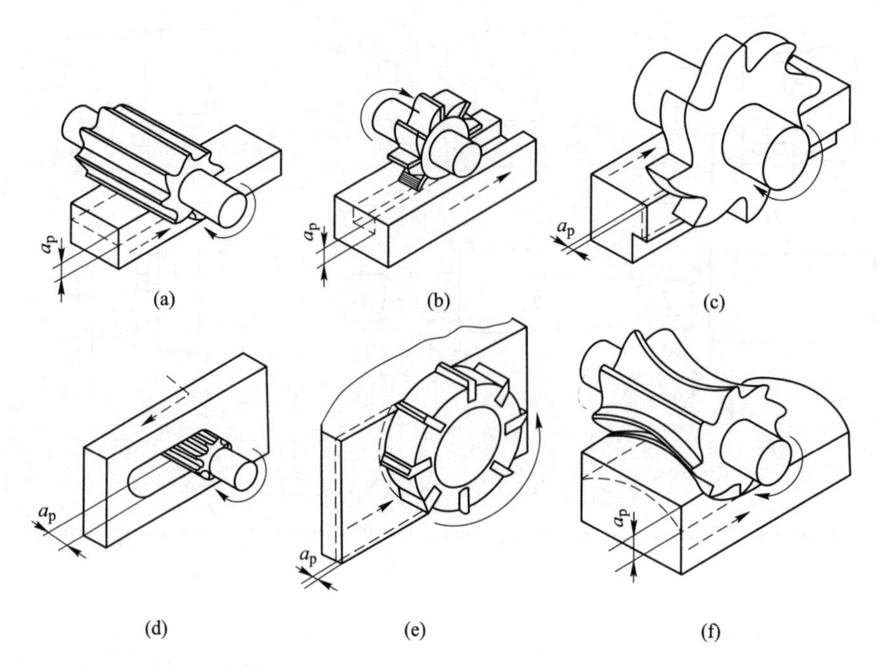

Figure 3　Some typical milling operations

(a) Slab or plain milling; (b) Slat or keyway milling; (c) Side milling; (d) End milling; (e) Face milling; (f) Form milling

Cutter rotation; - - -→ Feed motion

The Future of Manufacturing

Based on the study of how the manufacturing environment might appear in 2020 and beyond, performed by the National Research Council, USA, a number of key technologies were revealed that will have a great impact on the future of manufacturing. They are:

(1) Adaptable and reconfigurable manufacturing processes and systems.

(2) Extension of systems models for all manufacturing enterprise operations.

(3) Waste-free manufacturing.

(4) Synthesis and architecture technologies for converting information into desired knowledge.

(5) Processes and totally new customized materials with order of magnitude property improvements designed on the atomic scale (nanotechnology).

(6) Design methods and manufacturing processes for products than can easily be reconfigured with software or hardware.

(7) Direct machine/user interfaces that enhance human performance and promote intelligent input.

(8) Design methodologies that allow the process to enhance the range of product requirements by an order of magnitude.

(9) Biotechnology processes for manufacturing applications.

(10) Processes for rapid and cost-effective development, transfer and utilization of manufacturing technology.

During this decade, the most promising of the evolving technological methodologies are concurrent engineering (CE) and artificial intelligence (AI). On the other hand, the computer-integrated manufacturing (CIM) system concept and the formation of virtual companies are the most promising from the second group.

(Source: *Wit Grzesik. Advanced Machining Processes of Metallic Materials*. Amsterdam: Elsevier Science, 2017)

Words and Expressions

geometrically　*ad.* 几何学上的，按几何级数地
chip-forming　*n.* 切削形成
shearing　*n.* 剪切，剪断加工
blanking　*n.* 下料加工，冲裁（切）
punching　*n.* 冲剪，冲孔，冲压
blade　*n.* 刀片，刀刃
cutting　*n.* 切割，切削
turning　*n.* 车削
drilling　*n.* 钻削，钻孔
cylindrical　*a.* 圆柱形的，圆柱体的
taper　*n.* 锥形化，锥度
lathe　*n.* 车床；*v.* 用车床加工
workpiece　*n.* 工件，轧件
counterdrilling　*n.* 埋头钻
boring　*n.* 钻孔，扩洞，反钻
counterboring　*v.* 扩孔，方形扩孔
countersinking　*v.* 穿孔，锥形扩孔
chamfering　*n.* ［机］倒角；*v.* 切斜边
swarf　*n.* 削屑，金属的切屑
wedge-shaped　*a.* 楔形的，V 形的
tolerance　*n.* 界值，容差，公差
kinematical　*a.* 运动学的
prismatic　*a.* 棱柱的，菱形的，棱镜的
helical　*a.* 螺旋形的
sculpture　*n.* 雕饰，雕纹
cutter　*n.* 刀具，切割机，切割者
facing　*n.* 面车削，表面加工
profiling　*n.* 剖析
parting　*n.* 分段加工，分型
cutoff　*n.* 切角，切断，截断
threading　*n.* 穿线，喂料工序
slotting　*n.* 切缝切削，插削，开槽
longitudinal　*a.* 长度上，经向
reconfigurable　*a.* 可重构的
machining process　加工工艺
core drilling　取芯钻探，钻取岩芯
face milling　［机］面铣，端面铣削
peripheral milling　［机］圆周铣削，周缘铣
end milling　端铣，立铣

Phrases

deal with　处理，涉及
in terms of　依照，按照，在……方面
play a crucial role　起到关键作用
the most promising of　最有潜力的

Notes

(1) The term machining deals with any process in which material is removed gradually from a workpiece, including metal cutting with both single-point or multi-point tools (tools with geometrically defined cutting edges) and grinding with abrasive wheels which consists of a large

number of micro-cutting edges, randomly shaped and oriented, i. e. tools with geometrically undefined cutting edges.

in which 引导定语从句，先行词为 any process；后面的 which 引导的定语从句修饰 abrasive wheels。

参考译文：机械加工指的是从工件上逐渐去除边料的一些过程，包括使用单点或多点工具（几何上定义的刃口的工具）进行金属切削，以及使用由大量无规形状和定向的微削刃组成的砂轮进行磨削（几何不确定的切削刃的工具）。

(2) While metal cutting is commonly associated with big industries (automotive, aircraft, aerospace, home appliance, etc.) that manufacture big products, the machining of metals and alloys plays a crucial role in a range of manufacturing activities, including the ultraprecision machining of extremely delicate components.

associate with 译为"和……有关联"；play a crucial role 译为"起至关重要的作用"。

参考译文：虽然金属切削通常与制造大型部件的大工业（如汽车、飞机、航空航天、家电等）相关联，但金属和合金的加工在一系列制造活动中也发挥着至关重要的作用，包括对其极精细部件进行的超精密加工。

(3) Peripheral (plain or slab) milling generates a surface parallel to the axis of rotation while face milling generates a flat surface normal to the axis of rotation.

peripheral milling 译为"圆周铣削，周缘铣"；face milling 译为"面铣，端面铣削，平铣"。

参考译文：圆周（平面或平板）铣削产生一个平行于旋转轴的表面，而端面铣削产生一个平面垂直于旋转轴。

10.2　阅读材料

10.2　Reading Materials

Introduction to Machining Processes

Machining is a term that covers a large collection of manufacturing processes designed to remove unwanted material, usually in the form of chips from a workpiece. Machining is used to convert castings, forgings, or preformed blocks of metal into desired shapes, with size-and finish specified to fulfill design requirements. Almost every manufactured product has components that require machining, often to great precision. Machining processes are performed on a wide variety of machine tools. Examples of basic machine tools are milling machines, drill presses, grinders, shaper, broaching machines and saws.

The primary chip formation processes are listed below, with alternative versions in parentheses. Each process is performed on one or more of the basic machine tools. For example, drilling can be performed on drill presses, milling machines, lathes, and some boring machines:

(1) Turning (boring, facing, cutoff, taper turning, form cutting, chamfering recessing, thread cutting).

(2) Shaping (planing, vertical shaping).

(3) Milling (hobbing, generating, thread milling).

(4) Drilling (reaming, tapping, spot facing, counterboring, countersinking).

(5) Sawing (filing).

(6) Abrasive machining (grinding, honing, lapping).

(7) Broaching (internal and surface).

Processes can be combined into multiple-capability machine, known as machining centers. The machining center is capable of performing the machining processes normally performed on a milling machine, drilling machine and a boring mill and is numerically controlled.

1. Overview of Machining Process Variables

Metal cutting, processes can be viewed as consisting of independent (input) variables., dependent variables, and independent-dependent interactions or relationships. Figure 1 summarizes the input/output relationships associated with metal cutting. Understanding the connections between input variables and process behavior is important knowledge for the manufacturing engineer. Unfortunately, this knowledge is difficult to obtain. Machining is a unique plastic deformation process in that it is constrained only by the cutting tool and operates at very large strains and very high strain rates. The tremendous variety in the input variables results in an almost infinite number of different machining combinations. Basically, there are three ways to deal with such a complex situation.

(1) Experience

Experience requires long-term exposure, because knowledge is basically gained by trial and error, with successful combinations transferred to other "Similar" situations. This activity goes on in manufacturing every time a new material is introduced into the production facility. It took years for industry to learn how to machine titanium. Unfortunately, the knowledge gained through one process may not transfer well to another even though their input variables appear very similar.

(2) Experiments

Machining experiments are expensive, time consuming, and difficult to carry out. Tool life experiments, for example, are quite commonly done, yet tool life data for most workpiece/tool material combination are not available. Even when laboratory data have been published, the results are not necessarily transferable to the particular machine tools and cutting tools on the shop floor. Tool life equations are empirically developed from turning experiments in which all input variables except cutting speed are kept constant. The experimental arrangement may limit the mode of tool failure to wear. Such results are of little value on the shop floor, where tools can and do fail from causes other than wear.

(3) Theories

There have been many attempts to build mathematical models of the metal cutting process. These theories try to predict the direction of the shearing process of metal cutting. These models range

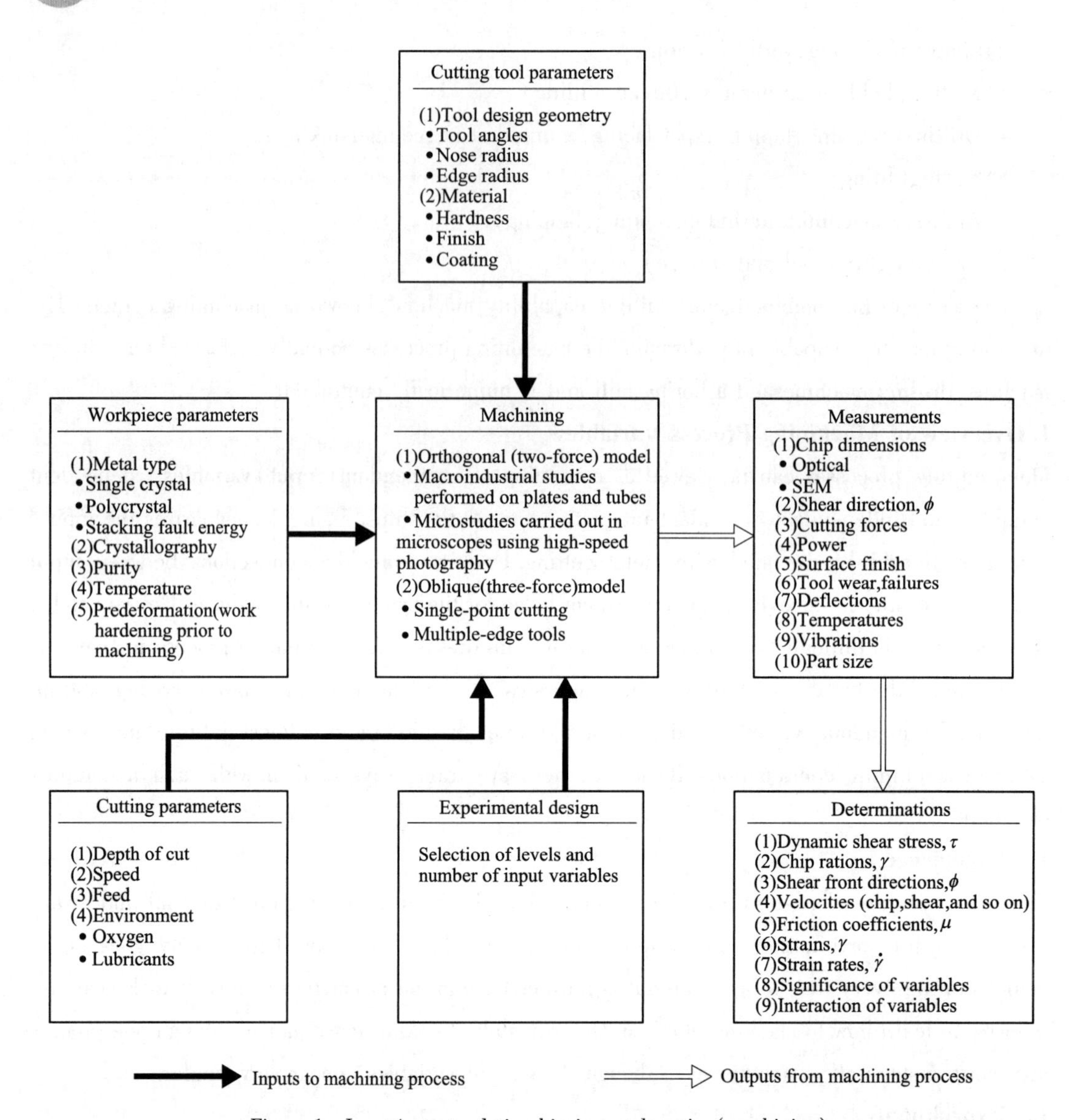

Figure 1　Input/output relationship in metal cutting(machining)

from crude, first-order approximation to complex, computer-based models using finite-element analysis. Recently, some modest successes have been reported in the literature in which accurate predictions of cutting forces and tool wear were made for certain materials. Clearly such efforts are extremely helpful in understanding how the process behaves. However, the theory of plastic deformation of metals (dislocation theory) has not yet been able to predict values for shear stresses and tool/chip interface from the metallurgy and deformation history of the material. Therefore, it has been necessary to devise two independent experiments to determine the shear strength (T_s) of the metal at large strains and high strain rates and the sliding fiction situation at the interface between the tool and chip.

2. Future Trends

The metal cutting process will continue to evolve, with improvements in cutting tool materials and machine tools leading the evolution. More refined coatings on cutting tools will improve tool life and reliability. The challenge for machining will involve dealing with the new types of materials that will need to be machined, including aluminum and titanium alloys, alloy, steels, and super-alloys. These materials, because of improved processing techniques, are becoming stronger and harder and therefore more difficult to machine. The objective should be to design and build culling tools that have less variability in their tool lives rather than longer tool lives. The increasing use of structural ceramics, high-strength polymers, composites, and electronic materials will also necessitate the use of nontraditional methods of machining. In addition, grinding will be employed to a greater extent than in the past, with greater attention to creep feed grinding and the use of superabrasives (diamond and cubic boron nitride).

As the cutting cools improve, the machine tools will become smarter, with on-board computers providing intelligent algorithms interacting with sensory data from the process. Programmable machine tools, if equipped with the proper sensor, are capable of carrying out measurements of the product as it is being produced. These product data will be fed back to the control program, which is then modified to improve the product or corrected for errors. Thus, the machine will be able to make the adjustments necessary to prevent defective prod-ucts from being produced. The goal of such control programs should be improved quality (designed not to make a defect), rather than optimum speed or lowest cost. Advancements in computer-aided machining processes are discussed in the Section "Machine Controls and Computer Applications in Machining" in this Volume.

Another area in which significant advances will be made is the design of workholders that are capable of holding various parts without any downtime for setups. Included in this search for flexible fixtures will be workholding devices that can be changed over by a robot—the same robot used to load or unload parts from the machine.

(Sonrce: J T Black. et al. *Introduction to Machining Processes*. ASM International Website)

Words and Expressions

broaching *n.* 拉孔，拉刀切削，扩孔
sawing *n.* 锯切，锯割
honing *n.* 珩磨
lapping *n.* 研磨，抛光，搭接
workholder *n.* ［机］工件夹具
flexible *a.* 灵活的，柔韧的，易弯曲的
polymer *n.* 聚合物
reaming *n.* 铰刀，纹孔
drill press ［机］钻床
milling machine ［机］铣床
thread milling 螺纹铣削
chamfering recessing 倒角凹槽
taper turning 锥度切削
spot facing 锪孔，局部整平

10.3　问题与讨论

10.3　Questions and Discussions

(1) The definition of machining.

(2) Classification of machining processes.

(3) The differences of turning, drilling and milling.

(4) What will have a great impact on the future of manufacturing?

(5) What are the most promising of the evolving technological methodologies?

单元 11　金属加工成型
Unit 11　Metal Working

扫码查看
讲课视频

11.1　教学内容
11.1　Teaching Materials

1. Introduction

Metalworking consists of deformation processes in which a metal billet or blank is shaped by tools or dies. The design and control of such processes depend on an understanding of the characteristics of the workpiece material, the conditions at the tool/workpiece interface, the mechanics of plastic deformation, the equipment used, and the finished-product requirements. These factors influence the selection of tool geometry and material as well as processing conditions (for example, workpiece and die temperatures and lubrication). Because of the complexity of many metalworking operations, models of various types, such as analytic, physical, or numerical models, are often relied upon to design such processes.

2. Historical Perspective

Metalworking is one of three major technologies used to fabricate metal products; the others are casting and powder metallurgy. However, metalworking is perhaps the oldest and most mature of the three. The earliest records of metalworking describe the simple hammering of gold and copper in various regions of the Middle East around 8000B.C. The forming of these metals was crude because the art of refining by smelting was unknown. With the advent of copper smelting around 4000B.C., a useful method became available for purifying metals through chemical reactions in the liquid state. Later, in the Copper Age, it was found that the hammering of metal brought about desirable increases in strength (a phenomenon now known as strain hardening). The quest for strength spurred a search for alloys that were inherently strong and led to the utilization of alloys of copper and tin (the Bronze Age) and iron and carbon (the Iron Age). The Iron Age, which can be dated as beginning around 1200 B.C., followed the beginning of the Bronze Age by some 1300 years. The reason for the delay was the absence of methods for achieving the high temperatures needed to melt and to refine iron ore.

Most metalworking was done by hand until the 13th century. At this time, the tilt hammer was developed and used primarily for forging bars and plates. This relatively simple device remained in service for some centuries. The development of rolling mills followed that of forging equipment. However, the development of large mills capable of hot rolling ferrous materials took almost 200 years. This relatively slow progress was primarily due to the limited supply of iron. During the Industrial Revolution at the end of the 18th century, processes were devised for making iron and steel in large quantities to satisfy the demand for metal products. A need arose for forging equipment with larger capacity. This need was answered with the invention of the high-speed steam hammer. From such equipment came products ranging from firearms to locomotive parts. Similarly, the steam engine spurred developments in rolling, and in the 19th century, a variety of steel products were rolled in significant quantities. The past 100 years have seen the development of new types of metalworking equipment and new materials with special properties and applications. The new types of equipment have included mechanical and screw presses and high-speed tandem rolling mills.

3. Classification of Metalworking Processes

In metalworking, an initially simple part—a billet or a blanked sheet, for example—is plastically deformed between tools (or dies) to obtain the desired final configuration. Metal-forming processes are usually classified according to two broad categories.

(1) Bulk, or massive, forming operations.

(2) Sheet forming operations.

In both types of process, the surfaces of the deforming metal and the tools are in contact, and friction between them may have a major influence on material flow. In bulk forming, the input material is in billet, rod, or slab form, and the surface-to-volume ratio in the formed part increases considerably under the action of largely compressive loading. In sheet forming, on the other hand, a piece of sheet metal is plastically deformed by tensile loads into a three-dimensional shape, often without significant changes in sheet thickness or surface characteristic.

Examples of generic bulk forming processes are extrusion, forging, rolling, and drawing. Specific bulk forming processes are listed in Table 1.

Table 1 Classification of bulk (massive) forming processes

Forging	Rolling	Extrusion	Drawing
Closed-die forging with flash Closed-die forging without flash Coining Electro-upsetting Forward extrusion forging Backward extrusion forging Hobbing Isothermal forging Nosing Open-die forging Rotary (orbital) forging Precision forging Metal powder forging Radial forging Upsetting	Sheet rolling Shape rolling Tube rolling Ring rolling Rotary tube piercing Gear rolling Roll forging Cross rolling Surface rolling Shear forming Tube reducing	Nonlubricated hot extrusion Lubricated direct hot extrusion Hydrostatic extrusion	Drawing Drawing with rolls Ironing Tube sinking

4. Types of Metalworking Equipment

The various forming processes discussed above are associated with a large variety of forming machines or equipment, including the following.

(1) Rolling mills for plate, strip, and shapes.

(2) Machines for profile rolling from strip.

(3) Ring-rolling machines.

(4) Thread-rolling and surface-rolling machines.

(5) Magnetic and explosive forming machines.

(6) Draw benches for tube and rod; wire- and rod-drawing machines.

(7) Machines for pressing-type operations.

Among those listed above, pressing-type machines are the most widely used and are applied to both bulk and sheet forming processes. The significant characteristics of pressing-type machines comprise all machine design and performance data that are pertinent to the economical use of the machine.

(Source: S L Semiatin. *Introduction to Bulk-Forming Processes*. ASM Handbook, Volume 14A, 2005)

Words and Expressions

billet *n.* 钢坯，坯料，坯锭
blank *a.* 空白的；*n.* 空白处；*v.* 成为空白
geometry *n.* 几何学，几何形状
lubrication *n.* 润滑
refining *n.* 精炼，提炼
purifying *n.* 净化，精制
hammering *n.* 锻造，锤击
hydraulic *a.* 液压的，水力的
ubiquitous *a.* 普遍存在的
phenomenon *n.* 现象，非凡的人（或事物）
extrusion *n.* 挤出，推出
upsetting *n.* 缩锻，镦锻
locomotive *n.* 机车，火车头；*a.* 移动的
coining *n.* 压印加工；*v.* 冲制，压模
hobbing *n.* 滚齿机，滚刀
nosing *n.* 级面突缘，级嘴，金属护沿
isothermal *a.* 等温的，等温线的
drawing *n.* 牵引，拖拽
Ironing *n.* 熨烫，熨平机
hydrostatic *a.* 流体静力学的
strain hardening 加工硬化，应变强化
hydraulic press 液压机，水压机
rolling mill 轧机
compressive loading 压缩载荷
sheet forming 板材成形
bulk forming （块材，块体）体积成形

Phrases

rely upon 依赖，依靠
be applied to 被应用到
be pertinent to 是……相关的
arise for 因为……而出现
with the advent of 随着……的出现
ranging from…to… 范围（幅度）从……到……

Notes

(1) The design and control of such processes depend on an understanding of the

characteristics of the workpiece material, the conditions at the tool/workpiece interface, the mechanics of plastic deformation, the equipment used, and the finished-product requirements.

谓语 depend on 后面跟着五个并列成分。

参考译文：这种加工过程的设计和控制依赖于对工件材料的特性、工具/工件界面处的状态、塑性变形的机理、所使用的设备以及成品要求的了解。

(2) The reason for the delay was the absence of methods for achieving the high temperatures needed to melt and to refine iron ore.

the absence of 译为"不存在，缺乏"; the reason for 译为"……的原因"。

参考译文：延迟的原因是缺乏实现高温熔融和精炼铁矿石所需的方法。

(3) In metalworking, an initially simple part—a billet or a blanked sheet, for example—is plastically deformed between tools (or dies) to obtain the desired final configuration.

blank 做名词指"（金属或木头的）坯件，坯料"，做动词有"切割"的意思。

参考译文：在金属加工中，最初简单的零件（例如，钢坯或坯料片）在工具（或模具）间通过塑性变形以获得最终想要的构造。

(4) In bulk forming, the input material is in billet, rod, or slab form, and the surface-to-volume ratio in the formed part increases considerably under the action of largely compressive loading.

bulk forming 译为"体积成型，块体成型，散装成型"; surface-to-volume ratio 译为"面容比，比表面积"。

参考译文：在块体成型中，输入材料为钢坯，棒材或板坯形式，并且在压缩载荷作用下，成型零件的表面体积比显著增加。

(5) In sheet forming, on the other hand, a piece of sheet metal is plastically deformed by tensile loads into a three-dimensional shape, often without significant changes in sheet thickness or surface characteristic.

sheet forming 跟 bulk forming 对比，前者为二维片材成型，后者为三维块体成型。

参考译文：另一方面，在片材成形中，金属片由于拉伸而塑性变形为三维形状，而在片材厚度或表面特性上没有显著变化。

11.2　阅读材料

11.2　Reading Materials

Introduction to Forging Methods

1. What is Forging?

Forging is a manufacturing process involving the shaping of a metal through hammering, pressing, or rolling (Figure 1). These compressive forces are delivered with a hammer or die. Forging is often categorized according to the temperature at which it is performed—cold, warm, or hot forging.

A wide range of metals can be forged. Typical metals used in forging include carbon steel, alloy steel, and stainless steel. Very soft metals such as aluminum, brass, and copper can also be forged. The forging process can produce parts with superb mechanical properties with minimum waste. The basic concept is that the original metal is plastically deformed to the desired geometric shape—giving it higher fatigue resistance and strength. The process is economically sound with the ability to mass produce parts, and achieve specific mechanical properties in the finished product.

Figure 1 Forging involves the shaping of metal through compressive forces such as hammering, pressing, or rolling

2. History of Forging

Forging has been practiced by smiths for thousands of years. At first, bronze and copper were the most common forged metals, in the Bronze Age: later, as the ability to control temperature and the process of smelting iron was discovered, iron became the primary forged metal. Traditional products include kitchenware, hardware, hand tools, and edged weapons. The Industrial Revolution allowed forging to become a more efficient, mass-production process. Since then, forging has evolved along with advances in equipment, robotics, electronic controls, and automation. Forging is now a worldwide industry with modern forging facilities producing high-quality metal parts in a vast array of sizes, shapes, materials, and finishes.

Forging is one of the oldest known metalworking processes. Traditionally, forging was performed by a smith using hammer and anvil, though introducing water power to the production and working of iron in the 12th century allowed the use of large trip hammers or power hammers that increased the amount and size of iron that could be produced and forged. The smithy or forge has evolved over centuries to become a facility with engineered processes, production equipment, tooling, raw materials and products to meet the demands of modern industry.

In modern times, industrial forging is done either with presses or with hammers powered by compressed air, electricity, hydraulics or steam. These hammers may have reciprocating weights in the thousands of pounds. Smaller power hammers, 500lb (230kg) or less reciprocating weight, and hydraulic presses are common in art smithies as well. Some steam hammers remain in use, but they became obsolete with the availability of the other, more convenient, power sources.

3. Advantages

Forging can produce a piece that is stronger than an equivalent cast or machined part. As the metal is shaped during the forging process, its internal grain deforms to follow the general shape of the part. As a result, the grain is continuous throughout the part, giving rise to a piece with improved strength characteristics.

Some metals may be forged cold and offer less machine stock and tighter tolerances. Low carbon and alloy steels can be cold forged. In the case of hot forging, a high temperature furnace (sometimes referred to as the forge) will be required to heat ingots or billets. Owing to the massiveness of large forging hammers and presses and the parts they can produce, as well as the dangers inherent in working with hot metal, a special building is frequently required to house the operation. In the case of drop forging operations, provisions must be made to absorb the shock and vibration generated by the hammer. Most forging operations will require the use of metal-forming dies, which must be precisely machined and carefully heat treated to correctly shape the work piece, as well as to withstand the tremendous forces involved.

4. Forging Methods

There are several forging methods with different capabilities and benefits. The more commonly used forging methods include the drop forging methods, as well as roll forging.

(1) Drop forging derives its name from the process of dropping a hammer onto the metal to mold it into the shape of the die. The die refers to the surfaces that come into contact with the metal. There are two types of drop forging—open-die and closed-die forging. Dies are typically flat in shape with some having distinctively shaped surfaces for specialized operations.

1) Open-die forging (smith forging).

Open-die forging is also known as smith forging. A hammer strikes and deforms a metal on a stationary anvil. In this type of forging, the metal is never completely confined in the dies—allowing it to flow except for the areas where it is in contact with the dies. It is the operator's responsibility to orient and position the metal to achieve the desired final shape. Open-die forging is suitable for simple and large parts, as well as customized metal components. Advantages of open-die forging: Better fatigue resistance and strength; Reduces chance of error and/or holes; Improves microstructure; Continuous grain flow; Finer grain size.

2) Closed-die forging (impression-die).

Closed-die forging is also known as impression-die forging. The metal is placed in a die and attached to an anvil. The hammer is dropped onto the metal, causing it to flow and fill the die cavities. The hammer is timed to come into contact with the metal in quick succession on a scale of milliseconds. Excess metal is pushed out from the die cavities, resulting in flash. The flash cools faster than the rest of the material, making it stronger than the metal in the die. After forging, the flash is removed. Advantages of closed-die forging: Produces parts up to 25 tons; Produces near net shapes that require only a small amount of finishing; Economic for heavy production.

(2) Roll forging consists of two cylindrical or semi-cylindrical horizontal rolls that deform a round or flat bar stock. This works to reduce its thickness and increase its length. This heated bar is inserted and passed between the two rolls—each containing one or more shaped grooves—and is progressively shaped as it is rolled through the machine. This process continues until the desired shape and size is achieved.

Advantages of automatic roll forging: Produces little to no material waste; Creates a favorable grain structure in the metal; Reduces the cross-sectional area of the metal; Produces taper ends.

(3) Press forging uses a slow, continuous pressure or force, instead of the impact used in drop-hammer forging (Figure 2). The slower ram travel means that the deformation reaches deeper, so that the entire volume of the metal is uniformly affected. Contrastingly, in drop-hammer forging, the deformation is often only at the surface level while the metal's interior stays somewhat undeformed. By controlling the compression rate in press forging, the internal strain can also be controlled.

Advantages of press forging: Economic for heavy production; Greater accuracy in tolerances within 0.01 ~ 0.02 inch; Dies have less draft allowing for better dimensional accuracy; Speed, pressure, and travel of the die are automatically controlled; Process automation is possible; Capacity of presses range from 500 ~ 9000 tons.

Figure 2 Press forging uses a slow, continuous pressure or force to shape metal uniformly, instead of the impact used in drop forging

(4) Upset forging is a manufacturing process that increases the diameter of the metal by compressing its length. Crank presses, a special high-speed machine, are used in upset forging processes. Crank presses are typically set on a horizontal plane to improve efficiency and the quick exchange of metal from one station to the next. Vertical crank presses or a hydraulic press are also options.

Advantages of upset forging: High production rate of up to 4500 parts per hour; Full automation is possible; Elimination of the forging draft and flash; Produces little to no waste.

(5) In automatic hot forging, mill-length steel bars are inserted into one end of the forging machine at room temperature, and hot forged products emerge from the other end. The bar is heated with high-power induction coils to a temperature ranging from 2190 ~ 2370℉ in under 60 seconds. The bar is descaled with rollers and shared into blanks. At this point, the metal is transferred through several forming stages that can be coupled with high-speed cold-forming operations. Typically, the cold-forming operation is left for the finishing stage. By doing so, the benefits of cold-working can be reaped while also maintaining the high speed of automatic hot forging.

Advantages of automatic hot forging: High output rate; Acceptance of low-cost materials; Minimal labor required to operate machinery; Produces little to no material waste (material savings between 20%~30% over conventional forging).

(6) Precision forging (net-shape or near-net-shape forging) requires little to no final machining. It is a forging method developed to minimize the cost and waste associated to post-forging operations. Cost savings are achieved from the reduction of material and energy, as well as the reduction of machining.

(7) Isothermal forging is a forging process where the metal and die are heated to the same temperature. Adiabatic heating is used—there is no net transfer of mass or thermal exchange between the system and the external environment. The changes are all due to internal changes resulting in highly controlled strain rates. Due to the lower heat loss, smaller machines may be used for this forging process.

(Source: *AnIntro to Forging Methods*. RELIANCE FOUNDRY Website)

Words and Expressions

die　*n.* 金属模具，压模
accuracy　*n.* 准确性，精确性
crank　*n.*（L字形）曲柄
coil　*n.*（绳或线等的）卷，匝，线圈
adiabatic　*a.*［物］绝热的，隔热的
roll forging　［机］滚锻
upset forging　［机］顶锻；平锻模
press forging　［机］冲锻
hot forging　［机］热锻
precision forging　［机］精密锻造

11.3 问题与讨论

11.3 Questions and Discussions

(1) What is metalworking?
(2) Introduce other technologies used to fabricate metal products.
(3) What spurred the developments in rolling?
(4) Classifications of metal-forming.
(5) Examples of generic bulk forming processes.

单元 12 轧制工艺介绍
Unit 12 Introduction to Rolling Process

扫码查看
讲课视频

12.1 教学内容
12.1 Teaching Materials

1. Definition of Rolling Process

Rolling is the most important metal forming process. More than 95% of ferrous and non-ferrous metals and alloys are processed to their usable shapes by rolling. Usable shapes of rolled metals are plate, sheet, strip, foil, different sections like rail, beam, channel, angle, bar, rod, and seamless pipe, etc., as shown in Figures 1 and 2.

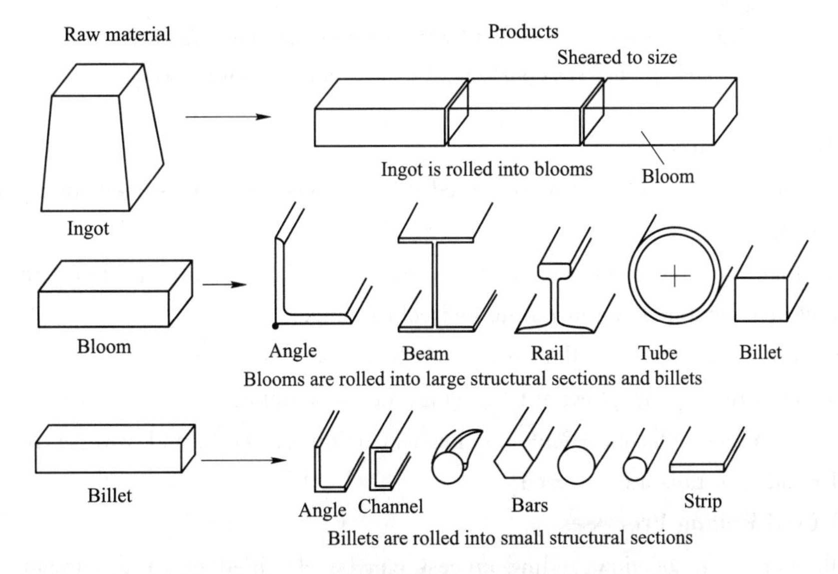

Figure 1 Products produced by hot rolling

In the rolling process, permanent deformation is achieved by subjecting the material to high compressive stress by allowing the material to pass through the gap between two rotating cylindrical rolls. The rolls may be flat or grooved, and are kept at a fixed distance apart from each other. The rolls are rotated in opposite direction by means of electrical drive system (motor, gearbox, spindle and couplings). Depending on the direction of rotation of the rolls, the input material enters the gap between the rolls from one end and comes out from the other end with a reduced cross-section, the roll gap area being kept less than the cross-sectional area of the input material. For obtaining the desired final shape of rolled material, it is generally necessary to pass the material through the

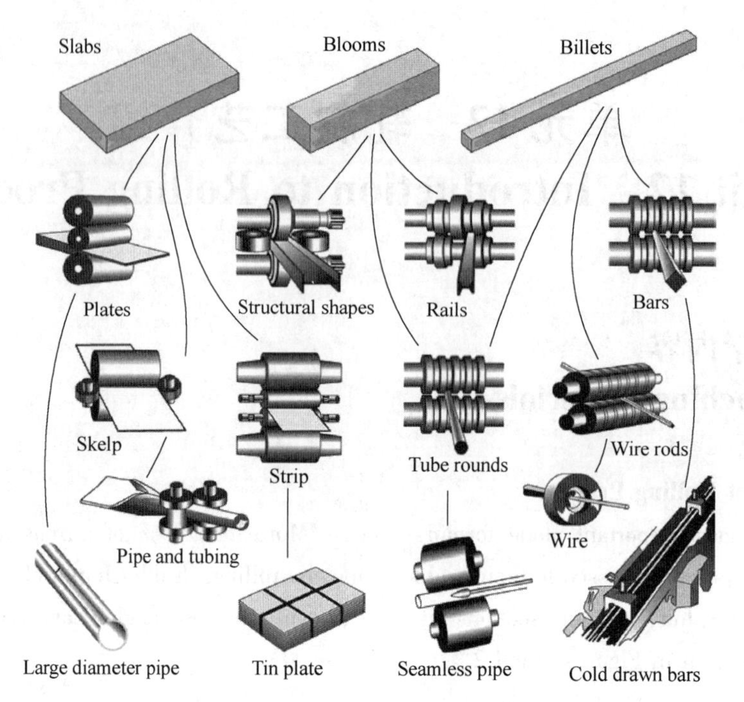

Figure 2 Schematic flowchart for the production of various finished and semi-finished steel products which pass through rolling process

rotating rolls several times.

Rolling process can be classified based on various conditions/methods employed in rolling. These are:

(1) Temperature of the material—thus we can have hot rolling (temperature above the recrystallization temperature), warm rolling and cold rolling.

(2) Shape of the rolled product-flat, sections or hollow sections rolling.

(3) Direction of rolling—lengthwise, transverse, and skew rolling.

(4) Mode of rolling mill operation—continuous (unidirectional), and reverse rolling, where direction of rotation of rolls are reversed.

2. Hot and Cold Rolling Processes

From metallurgical point of view, rolling process can be classified under two broad categories, namely (1) hot rolling and (2) cold rolling.

(1) Hot rolling

In hot rolling the material is rolled at a temperature higher than its recrystallization temperature. The advantage of hot rolling is twofold. First, at elevated temperature the strength of any metal or alloy is reduced. Thus the compressive force required for deformation is comparatively less and therefore smaller capacity rolling stand can be used for rolling operation. The second advantage of rolling a material at a temperature higher than its recrystallization temperature is that a large amount of plastic deformation can be imparted without getting it strain hardened.

The ferrous raw material for rolling various shapes is the ingot which is cast out of molten metal. In case of low carbon steels the ingot is quite large. It is first rolled into blooms. The blooms are rolled into smaller sizes, called billets. Large structural sections such as rails, beams, girders, channels, angle sections, and plates are rolled out of blooms, while billets are rolled into smaller structural sections, bars, plates, and strips. Rolling of ingots to blooms and blooms to billets, and further rolling of blooms and billets to different usable products like structural sections, bars, plates, and strips are all rolled through hot rolling. Some of the products produced through hot rolling process are illustrated in Figure 1.

(2) Cold Rolling

When rolling of a material is done at room temperature or below the recrystallization temperature of the material, it is called cold rolling. Obviously, the advantages of hot rolling is absent in cold rolling. The resistance to deformation is more. Furthermore, during rolling, strain hardening takes place, i. e. , the strength of the material progressively increases with increase in degree of deformation in the original material. However, there are a few advantages also. The first one is about controlling the grain size and thereby achieving the desired mechanical properties of the finished rolled material. When the input material is cold-rolled, the grains of the input material get elongated along the direction of rolling. Thus the effective grain size is reduced, as the surface area of each grain increases whereas their volume remains the same. With subsequent passes of rolling, the elongated grains break and the grain size becomes progressively smaller and the material gets harder and harder. After a certain percentage of volumetric deformation, the cold-rolled material becomes too hard and brittle to be rolled further profitability. At this stage, the cold-rolled material is annealed, which is nothing but heating the material in a neutral atmosphere above its recrystallization temperature. By adjusting the time for which the rolled material is kept at this higher temperature (soaking time), the size of the newly formed grains of the annealed material can be closely controlled.

Cold rolling is generally done to produce flat rolled products like sheet, plate, strip, and foil. In cold rolling, since the degree of deformation, i. e. , reduction in thickness of the flat product in any rolling pass, is kept low to avoid high roll separating force, several rolling passes are generally required along with requisite number of intermediate annealing. Furthermore, various advanced techniques and systems are employed to keep the rolled material flat and the thickness of the finished product within close tolerance throughout the length and width of the product.

(Source: Siddhartha Ray. *Principles and Applications of Meta Rolling*. Cambridge: Cambridge University Press, 2016)

Words and Expressions

strip n.（纸、金属、织物等）条，带；带状地带
motor n. 发动机，马达，汽车
gearbox n. 变速箱，齿轮箱
spindle n. 主轴，纺锤，锭子
coupling n.［电］耦合，联结，联轴器
lengthwise a. 纵向的
transverse a. 横向的；横断的
skew a. 歪斜的，［数］异面的
recrystallization n. 再结晶

bloom *n.* 花朵，方坯
girder *n.* [建] 大梁，纵梁
elongated *v.* 把……拉长，延长；*a.* 细长的
groove *n.* 有沟的，凹槽
volumetric *a.* 容积（量）的
profitability *n.* 盈利能力，收益性，利益率
neutral *a.* 中立的，中性的，不带电的
permanent deformation 永久变形
seamless pipe 无缝管
input material 原料，进料
rolling stand 轧钢机架
plastic deformation 塑性变形
warm rolling 温轧
cold rolling 冷轧，冷压延
soaking time 均热时间，保温时间，浸渍时间

Phrases

from…point of view 从……角度看
by means of 通过……方法
be absent in 不存在

Notes

(1) In the rolling process, permanent deformation is achieved by subjecting the material to high compressive stress by allowing the material to pass through the gap between two rotating cylindrical rolls.

achieved by 译为"通过……达到"；subject to 译为"使遭受"；allow to 译为"使允许，同意"；pass through 译为"穿过，通过"。

参考译文：在轧制过程中，通过使材料穿过两个旋转的圆柱辊之间的间隙，让材料承受高压缩应力，从而实现永久变形。

(2) Rolling of ingots to blooms and blooms to billets, and further rolling of blooms and billets to different usable products like structural sections, bars, plates, and strips are all rolled through hot rolling.

ingot 译为"铸锭，铸块"；hot rolling 译为"热轧（工艺）"。

参考译文：将钢锭轧制成大方坯，将大方坯轧制成小方坯，再将大方坯和小方坯进一步轧制成不同的可用产品，如结构型材、棒材，板坯和带材、均可通过热轧进行轧制。

(3) When the input material is cold-rolled, the grains of the input material get elongated along the direction of rolling. Thus the effective grain size is reduced, as the surface area of each grain increases whereas their volume remains the same.

cold rolling 译为"冷轧"；input material 译为"原料"；effective 在此表示"实际的，有效的"。

参考译文：当原料被冷轧时，原料的晶粒沿轧制方向拉长。由于每个晶粒的表面积增加，而它们的体积保持不变。因此，实际的晶粒尺寸减小。

(4) In cold rolling, since the degree of deformation, i. e., reduction in thickness of the flat product in any rolling pass, is kept low to avoid high roll separating force, several rolling passes are generally required along with requisite number of intermediate annealing.

the degree of 译为"……的程度"；i. e. 译为"即……"，此处为插入语，实际句子的主语为 several rolling passes。

参考译文：在冷轧中，由于变形程度保持较低，即在任何轧制道次中平板产品厚度的减小，以避免高的轧辊分离力，通常需要多次轧制道次以及必要的中间退火。

12.2　阅读材料
12.2　Reading Materials

Introduction to welding process

1. Introduction

Welding is a process in which two or more parts are joined permanently at their touching surfaces by a suitable application of heat and/or pressure. Often a filler material is added to facilitate coalescence. The assembled parts that are joined by welding are called a weldment. Welding is primarily used in metal parts and their alloys.

Welding processes are classified into two major groups:

(1) Fusion welding: In this process, base metal is melted by means of heat. Often, in fusion welding operations, a filler metal is added to the molten pool to facilitate the process and provide bulk and strength to the joint. Commonly used fusion welding processes are: arc welding, resistance welding, oxyfuel welding, electron beam welding and laser beam welding.

(2) Solid-state welding: In this process, joining of parts takes place by application of pressure alone or a combination of heat and pressure. No filler metal is used. Commonly used solid-state welding processes are: diffusion welding, friction welding, ultrasonic welding.

2. Arc Welding and Similar Processes

Arc welding is a method of permanently joining two or more metal parts. It consists of combination of different welding processes wherein coalescence is produced by heating with an electric arc, (mostly without the application of pressure) and with or without the use of filler metals depending upon the base plate thickness. A homogeneous joint is achieved by melting and fusing the adjacent portions of the separate parts. The final welded joint has unit strength approximately equal to that of the base material. The arc temperature is maintained approximately 4400℃. A flux material is used to prevent oxidation, which decomposes under the heat of welding and releases a gas that shields the arc and the hot metal.

3. Shielded-Metal Arc (SMAW) or Stick Welding

This is an arc welding process wherein coalescence is produced by heating the workpiece with an electric arc setup between a flux-coated electrode and the workpiece. The electrode is in a rod form coated with flux. Figure 1 illustrates the process.

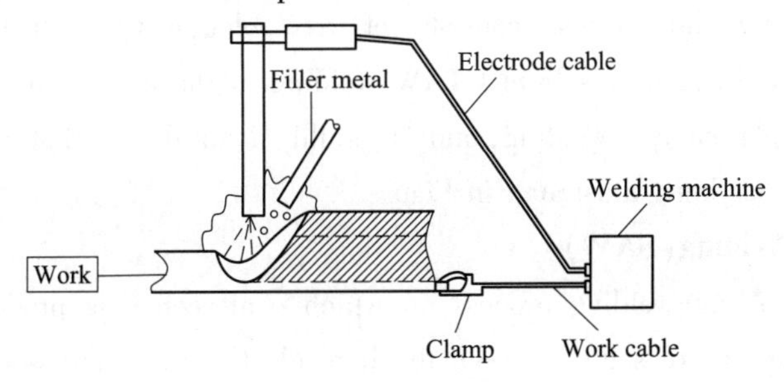

Figure 1　Shielded-Metal Arc (SMAW)

4. Flux-Cored Arc Welding (FCAW)

This process is similar to the shielded-arc stick welding process with the main difference being the flux is inside the welding rod. Tubular, coiled and continuously fed electrode containing flux inside the electrode is used, thereby, saving the cost of changing the welding. Sometimes, externally supplied gas is used to assist in shielding the arc.

5. Gas-Metal Arc Welding (GMAW)

In this process an inert gas such as argon, helium, carbon dioxide or a mixture of them are used to prevent atmospheric contamination of the weld. The shielding gas is allowed to flow through the weld gun. The electrode used here is in a wire form, fed continuously at a fixed rate. The wire is consumed during the process and thereby provides filler metal. This process is illustrated in Figure 2.

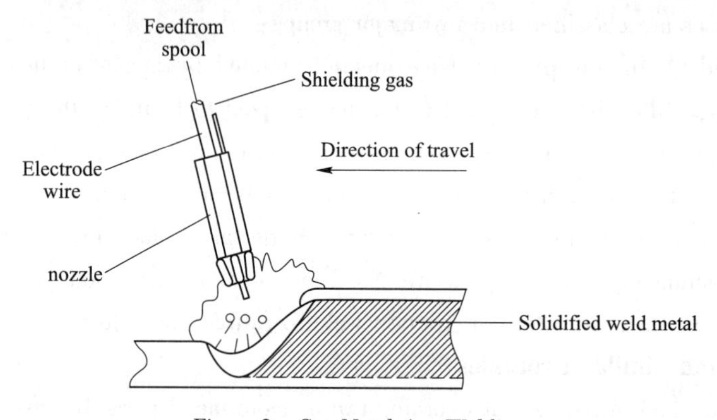

Figure 2 Gas-Metal Arc Welding

6. Gas-Tungsten Arc Welding (GTAW)

This process is also known as tungsten-inert gas (TIG) welding. This is similar to the Gas-Metal Arc Welding process. Difference being the electrode is non consumable and does not provide filler metal in this case. A gas shield (usually inert gas) is used as in the GMAW process. If the filler metal is required, an auxiliary rod is used.

7. Plasma Arc Welding (PAW)

This process is similar to TIG. A non-consumable electrode is used in this process. Arc plasma is a temporary state of gas. The gas gets ionized after the passage of electric current and becomes a conductor of electricity. The plasma consists of free electrons, positive ions, and neutral particles. Plasma arc welding differs from GTAW welding in the amount of ionized gas which is greatly increased in plasma arc welding, and it is this ionized gas that provides the heat of welding. This process has been illustrated in Figure 3.

8. Submerged Arc Welding (SAW)

This is another type of arc welding process, in which coalescence is produced by heating the workpiece with an electric arc setup between the bare electrode and the work piece. Molten pool remains completely hidden under a blanket of granular material called flux. The electrode is in a

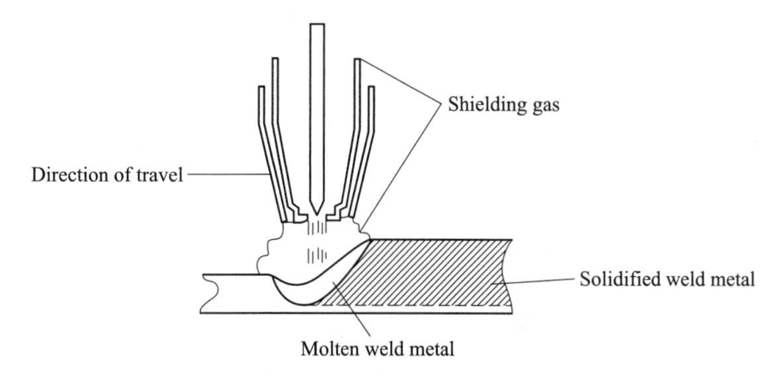

Figure 3 Plasma Arc Welding(PAW)

wire form and is continuously fed from a reel. Movement of the weld gun, dispensing of the flux and picking up of surplus flux granules behind the gun are usually automatic.

9. Oxyfuel Gas Welding(OGW)

This process is also known as oxy-acetylene welding. Heat is supplied by the combustion of acetylene in a stream of oxygen. Both gases are supplied to the torch through flexible hoses. Heat from this torch is lower and far less concentrated than that from an electric arc.

10. Resistance Welding

Resistance welding is a group of welding process in which coalescence is produced by the heat obtained from the resistance of the work to the flow of electric current in a circuit of which the work is a part and by the application of pressure. No filler metal is needed in this process.

11. Electron-Beam Welding(EBW)

Electron beam welding is defined as a fusion welding process wherein coalescence is produced by the heat obtained from a concentrated beam of high velocity electron. When high velocity electrons strike the workpiece, kinetic energy is transformed into thermal energy causing localized heating and melting of the weld metal. The electron beam generation takes place in a vacuum, and the process works best when the entire operation and the workpiece are also in a high vacuum of 10^{-4} torr or lower. However, radiations namely—ray, infrared and ultraviolet radiation generates and the welding operator must be protected.

12. Laser Beam Welding(LBW)

Laser beam welding is defined as a fusion welding process and coalescence is achieved by utilizing the heat obtained from a concentrated coherent light beam and impinging upon the surface to be joined. This process uses the energy in an extremely concentrated beam of coherent, monochromatic light to melt the weld metal. This process is illustrated in Figure 4.

13. Friction Welding(FRW)

In friction welding (solid state welding process) coalescence is produced by utilizing the heat obtained from the mechanically induced rotating motion between the rubbing surfaces. When the temperature at the interface of the two parts is sufficiently high, the rotation is stopped and

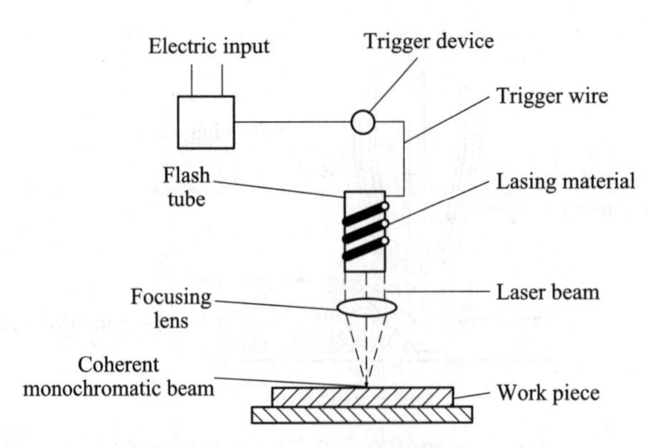

Figure 4 Laser-beam welding

increased axial force is applied. This fuses the two parts together. The rotational force is provided through a strong motor or a flywheel. In the latter case the process may be called inertia welding.

14. Other Welding Processes

Other processes used in the industry are following:

(1) Diffusion bonding(DB): Parts are pressed together at an elevated temperature below the melting point for a period of time.

(2) Explosion welding(EXW): The parts to be welded are driven together at an angle by means of an explosive charge and fuse together from the friction of the impact.

(3) Ultrasonic welding(USW) for metals: This process utilizes transverse oscillation of one part against the other to develop sufficient frictional heat for fusion to occur.

(4) Electro slag(ESW) and Electro gas(EGW) processes: In these processes a molten pool of weld metal contained by copper "shoes" is used to make vertical butt welds in heavy plate.

(Source: Herlin Puscan Jara. *Introduction to Welding Process*. Universidad Tecnologica del Peru, 2019)

Words and Expressions

coalescence *n.* 合并，联合，聚结
weldment *n.* 焊件，焊成件
monochromatic *a.* 单色的
argon *n.* [化] 氩 (Ar)
infrared *a.* 红外线的，使用红外的
ultraviolet *a.* 紫外的，紫外线的
molten pool 熔池
laser beam 激光
inert gas 惰性气体
plasma arc welding 等离子弧焊
positive ion 正离子
butt weld 对接焊缝，对头焊接
surplus flux 剩余磁通

12.3 问题与讨论

12.3 Questions and Discussions

(1) What kind of shapes can be formed by rolling?

(2) Based on the working temperature, how to classify rolling process?

(3) What is the working temperature of hot rolling?

(4) The advantages of hot rolling.

(5) What will happen during cold rolling?

(6) Why the cold-rolled metals should be annealed?

单元 13 研磨与抛光
Unit 13 Grinding and Polishing

扫码查看
讲课视频

13.1 教学内容
13.1 Teaching Materials

Grinding removes saw marks and levels and cleans the specimen surface. Polishing removes the artifacts of grinding but very little stock. Grinding uses fixed abrasives—the abrasive particles are bonded to the paper or platen—for fast stock removal. Polishing uses free abrasives on a cloth; that is, the abrasive particles are suspended in a lubricant and can roll or slide across the cloth and specimen. Some companies do not distinguish between grinding and polishing. Table 1 shows a typical ceramographic grinding and polishing procedure for an automatic polishing machine.

Table 1 A typical ceramographic grinding and polishing procedure for an automatic polishing machine

Step	Abrasive and lubricant	Time/min	Platen frequency /rpm	Head frequency /rpm
1. Plane grinding	240-grit bonded diamond disc sprayed continuously with water	0.5 ~ 1 (or until specimen is flat and saw marks are removed)	200~300	120~150 opposite to platen
2. Coarse polishing	15μm diamond suspended in water-soluble oil, sprayed every 20~30s on napless paper	5~10	120~150	120~150 opposite to platen
3. Polishing	6μm diamond suspended in water-soluble oil, sprayed every 20~30s on napless paper	5~10	120~150	120~150 opposite to platen
4. Fine polishing	1μm diamond suspended in water-soluble oil, sprayed every 20~30s on napless paper	5~10	120~150	120~150 opposite to platen
5(a). Relief polishing (optional)	0.05μm γ-Al_2O_3 slurry sprayed every 20~30s on napped cloth	1~5	120~150	120~150 opposite to platen
5(b). Vibratory polishing (optional)	Colloidal silica slurry, replenished every 30~60min on napped cloth	60~480	…	…

Note: For machines without timed spraying, the slurries can be poured from squeeze bottles or aerosols, or diamond pastes instead.

1. Automatic Grinding

The pressure, time, and starting abrasive size depend on the number ofmounts being ground, the abrasion resistance of the ceramic, the amount of wear on the abrasive particles, and the smoothness of the as-sawed surface. An automatic grinding and polishing machine is shown in Figure 1.

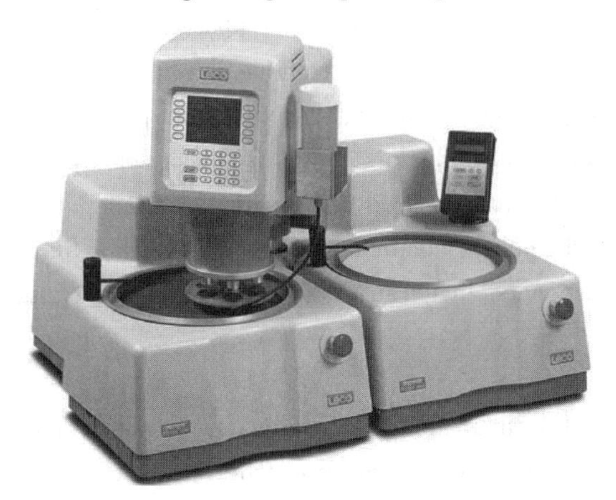

Figure 1 Automatic grinding and polishing machine

Automatic grinding method steps are:

(1) Symmetrically load three to six mounted specimens into the specimen holder of an automatic grinding-polishing machine, with the flat surface of the ceramic section downward.

(2) Grind the specimens at a contact pressure of 40 to 150kPa on a bonded diamond platen for approximately 60s or until the exposed surface of each specimen is flat and clean.

(3) Remove the specimen holder from the machine and clean the specimens. Once clean, return the specimen holder to the machine for polishing or more grinding in successive steps on ever-finer abrasives and follow each step with thorough cleaning.

2. Automatic Polishing

After the finest grinding step (1) in Table 1, polish the specimens on napless polishing cloths loaded with lubricant and progressively smaller diamond abrasives. Diamond polishing abrasives are typically available in 30, 15, 9, 6, 3, 1, and 0.25μm sizes, in liquid suspensions, pastes, and aerosols. The suspensions can be automatically sprayed by some machines at timed intervals. The transition from grinding to polishing may require additional time on the coarse polishing step (2) in Table 1 to remove the artifacts of grinding. If paste is used, reapply it to the polishing cloth every few minutes. All types of diamond abrasives break down quickly and should be replenished frequently.

Use napless cloth for diamond pastes or suspensions and napped cloth for the alumina slurry or colloidal silica. Napless cloth is a stiff, nonwoven PVC chemotextile sold under such trade names as Texmet, Pellon, DP-Plan, MD-Plan, and Pan-W. Nonwoven, fiber-reinforced-resin perforated pads and woven silk also work well for polishing ceramics with diamond pastes and suspensions. Flocked twill or napped cloth has a fuzzy texture that conforms to the surface being polished. Polishing lubricants come under various names, including lapping oil, diamond extender, and blue lubricant.

Step 5(a) in Table 1, relief polishing, is optional. Relief polishing is not recommended when the specimen is to be tested for microhardness. Relief polishing in conjunction with Nomarski differential interference contrast can enhance the contrast at low magnification by means of differential abrasion rates between harder and softer phases. Relief polishing can also polish the metal components in cross sections of microelectronic devices. Vibratory polishing with colloidal silica or alumina slurry, step 5(b) in Table 1, is another final polish technique. This method works very well for soft metals and semiconductors and is useful for some harder metals and ceramics.

In some cases, a corrosive liquid is used along with the relief polishing slurry in a technique called attack polish. Attack polish combines mild etching and final polishing into a single step. Colloidal silica is suspended in a caustic solution that has an attack-polish effect on some materials.

3. Manual Grinding

The manual method is useful when automatic equipment is not available or when the depth of grinding is critical. Cross sections of microelectronic devices, such as multiplayer packages, often must be ground to a specific depth. To grind aceramographic section manually, choose a reference point on the specimen, such as point Q in the 12 o'clock position shown in Figure 2(a). Hold the specimen surface firmly against the abrasive disc or belt such that the reference point is fixed with respect to the direction of abrasive motion. Continue grinding until the saw marks are replaced by the parallel scratches of the first abrasive, as in Figure 2(b). Rotate the reference point Q to the 3 o'clock position, as in Figure 2(c), and grind the specimen on the next finer abrasive until the previous artifacts are removed. The new parallel scratches lie at a 90° angle to the previous ones, as in Figure 2(d). Rotation of the mount by 90° after each abrasive step [Figure 2(e)] allows one to easily see when the artifacts of the previous preparation step have been removed. Clean the mount thoroughly after each step to prevent transfer of abrasive particles from one platen to the next.

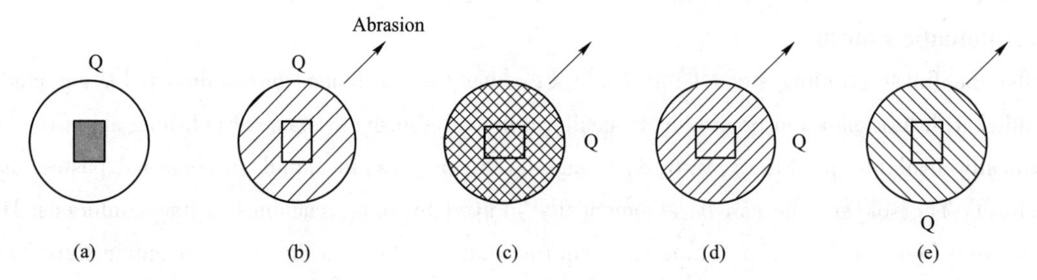

Figure 2 Sequence of steps (a) ~ (e) in manual grinding and polishing

4. Manual Polishing

After the finest grinding step and subsequent cleaning, manually polish the specimen on napless polishing cloths loaded with lubricant and 15, 6, and 1μm diamond paste, respectively. Rotate the specimen 90°, as in Figure 2(a) ~ (e), and clean it thoroughly after each polishing step. The relief polishing step with 0.05μm γ-Al_2O_3 suspension is optional.

(Source: *Grinding and Polishing*. ASM International Website)

Words and Expressions

specimen　*n.* 样品，样本
polishing　*n.* 磨光，抛光
artifact　*n.* 人工制品，历史文物
ceramographic　*a.* 陶瓷相的
abrasive　*n.* 研磨料
mount　*n.* 底座，底架
lubricant　*n.* 润滑剂，润滑油
suspension　*n.* 悬浮液，悬胶体
napless　*a.* 无绒的
nonwoven　*n.* 无纺布
perforated　*a.* 穿孔的，打孔的，有孔的
holder　*n.* 夹具，支架
symmetrically　*ad.* 对称地，匀称地
microhardness　*n.* 显微硬度
etching　*n.* 蚀刻术；v. 蚀刻；*a.* 蚀刻的
coarse　*a.* 粗糙的
replenish　*v.* 补充，重新装满
caustic　*a.* 腐蚀性的，苛性的
aerosol　*n.* 气凝胶
alumina　*n.* 氧化铝，矾土
colloidal　*a.* 胶体的，胶质的，胶状的
silica　*n.* 二氧化硅
scratch　*n.* 划痕，刮痕
perforated pads　多孔板，冲孔板
relief polishing　显微切片浮雕抛光
attack polish　腐蚀磨光法，侵蚀抛光法
flocked twill　植绒斜纹布
vibratory polishing　振动抛光

Phrases

be bonded to　黏着，连着
in conjunction with　连同……；与……一起
with respect to　关于，至于
be replaced by　由……所代替

Notes

(1) The pressure, time, and starting abrasive size depend on the number of mounts being ground, the abrasion resistance of the ceramic, the amount of wear on the abrasive particles, and the smoothness of the as-sawed surface.

depend on 后面为四个并列结构；abrasion resistance 译为“耐磨性”。

参考译文：压力、时间和起始磨料尺寸取决于用于研磨的底座量，陶瓷的耐磨性，磨料颗粒的磨损量以及锯齿表面的光滑度。

(2) Remove the specimen holder from the machine and clean the specimens. Once clean, return the specimen holder to the machine for polishing or more grinding in successive steps on ever-finer abrasives and follow each step with thorough cleaning.

successive step 译为“连续的步骤”；ever-finer 译为“更精细”。

参考译文：从机器上取下样品夹具并清洗样品。清洁后，将样品夹具放回机器进行抛光，或者在更精细的磨料中继续研磨，并在每个步骤后彻底清洗。

(3) After the finest grinding step (1), polish the specimens on napless polishing cloths loaded with lubricant and progressively smaller diamond abrasives.

loaded with 译为“负载有……”，“装载有……”。

参考译文：经过最细的研磨步骤（1）之后，在装入润滑剂和较小的金刚石磨料的无绒布上抛光样品。

（4）Relief polishing in conjunction with Nomarski differential interference contrast can enhance the contrast at low magnification by means of differential abrasion rates between harder and softer phases.

relief polishing 译为“显微切片浮雕抛光”；Nomarski differential interference contrast 是一种 Nomarski 差分干涉差显微镜术；in conjunction with 译为“连同，结合”。

参考译文：浮雕抛光结合 Nomarski 差分干涉差，可以通过硬相和软相间磨耗率的差异来提高低倍率下的对比度。

（5）Rotation of the mount by 90° after each abrasive step [Figure 2(e)] allows one to easily see when the artifacts of the previous preparation step have been removed.

when 引导条件状语从句。

参考译文：在每个研磨步骤［见图 2(e)］后将底座旋转 90°可以轻松地看到上一个制备步骤产生工件何时被移除。

13.2　阅读材料
13.2　Reading Materials

Abrasion

1. General Considerations

Abrasion is the most common type of wear; its destructive effects are widespread throughout most industries, including the pulp and paper industry. Abrasion is due to hard particles or protrusions being forced against, and moving along, a solid surface. It becomes more severe the harder the particles are, compared to the surface they are moving against. This chapter gives an overview of abrasion processes, and briefly reviews guidelines for materials selection, with a focus on stainless steels compared to other ferrous alloys. References [1] and [2] give a more thorough treatment of abrasion mechanisms. References [3] and [4] give useful information on the various options available for managing abrasion. Of the other types of wear that occur in the pulp and paper industry, erosion is also commonly encountered, due to the necessity of handling large volumes of process liquids through pipes and tanks. Unlike abrasion, erosive wear requires the flow of a liquid or gas stream past a surface. Its effects can be made worse if the stream contains abrasive particles. For materials options for controlling erosion, see Reference [3].

Abrasion is often classified as being low-stress, high-stress, or gouging abrasion. These are qualitative descriptors, but they can be useful for a first characterization of an abrasive situation. In low-stress abrasion, the amount of plastic deformation and work hardening of the abraded surface would be expected to be relatively low; damage occurs mainly by cutting and the abrading particles generally remain intact. In high-stress abrasion, imposed loads are greater, so that a higher proportion of surface plastic deformation and work hardening may be expected. At the same time, the amount of surface cutting also increases due to the creation of fresh abradant by fracture of existing abrading particles. Plastic deformation and work hardening predominate in gouging abrasion, where imposed loads are highest of all.

2. Materials Selection

Abrasion rates depend on many metallurgical and environmental factors. Therefore, it is important to realize that materials do not have an intrinsic level of abrasion resistance. Metallurgical factors that influence abrasion include hardness, microstructure (e. g., the presence of ferrite, austenite, pearlite or martensite in steels), alloy content, and the presence of second phase particles in the microstructure (e. g. carbides). Environmental factors include the nature of the abrasive (e. g. size, hardness, angularity) the presence of a corrosive liquid or gas, the temperature, load, sliding speed and humidity. For more details about the relative importance of these factors, see Reference [1].

Because of the complexity of abrasion processes, materials selection is not an exact science and, for every example of a successful application, there often seems to be a counter example. [1] Having said this, it is possible to provide some general guidelines.

For situations where corrosion is not a factor, the abrasion resistance of steels and cast irons will be largely dependent on their hardness, carbon content and microstructure. Generally, increased hardness and carbon content give increased levels of abrasion resistance, through the formation of martensite and carbides in the microstructure. [5] At the high end of the carbon range (e. g., with alloys such as the white cast irons), factors such as toughness and impact resistance become increasingly important. While hardness is a readily available guide to abrasion resistance, it must be remembered that there are many exceptions to the rule. This is because the true measure of abrasive wear resistance is the maximum hardness of the work hardened, abraded surface, and not the initial bulk hardness. Alloys that rapidly work harden as a result of abrasion can thus also give good performance.

This helps explain why austenitic stainless steels, which readily work harden but do not contain either martensite or carbides, can give acceptable performance in some abrasive wear conditions. [6] Increasing alloy content in austenitic stainless steels decreases their work hardening rate. [7] Type 304L (UNS S30403) and the proprietary alloy Nitronic 30 (UNS S20400) have been shown to resist some abrasive conditions better than Type 316L (UNS S31603), because of their higher work hardening. Duplex stainless steels, such as 2205 (UNS S32205), with their combination of work hardening capacity and improved strength and toughness, can be expected to outperform the austenitic stainless steels in terms of abrasion resistance. For the mill engineer, the practical problem is how to predict if a given abrasion situation will cause a sufficient amount of work hardening. In this respect, prior field experience is probably the best guide. [8]

In the pulp and paper industry, most abrasion problems will involve at least some amount of corrosion, and it is important to remember that the presence of a corrosive environment can significantly change the overall abrasion performance of many ferrous materials. This is especially true for alloy steels and white cast irons, which are often the first choice for combating "pure" abrasion. [9] Examples of mill equipment where a combination of abrasion and corrosion can occur include beaters, deflakers, stock refiners, chip refiners, pumps, storage bins, and chutes. In beaters, deflakers and stock refiners, abrasion-resistant cast irons [25% Cr and Ni-HiCr Type D (ASTM A532)] and precipitation hardening stainless steels [such as 17-4PH (UNS S17400)] are used in situations where abrasion resistance is the overriding consideration. The 17-4PH alloy would normally only be used for light abrasion duty; 25% Cr and Ni-HiCr Type D would be selected for more abrasive service. When both abrasion and corrosion are significant, the 25% Cr white cast irons or high carbon cast stainless steels are used, e. g., for refiner plates. In pumps, abrasion from sand and grit has led to widespread replacement of the standard CF-3 (UNS J92500) and CF-3M (UNS J92800) grades with the cast duplex grades, CD-6MN (UNS J93371) and CD-4MCuN (UNS J93372). These two cast duplex grades have much improved resistance to sand and grit erosion as well as better corrosion resistance. For storage bins and chutes, where a smooth surface must be maintained to allow easy flow of bulk solids, Types 409 (UNS S40900) and 304 (UNS S30400)[10] have replaced abrasion resistant ("AR") carbon steels, which rapidly corrode in mill process waters.[11] Duplex stainless steels are also increasingly used for these applications.

In each case, it is necessary to balance the amount of abrasive wear resistance needed with the appropriate level of corrosion resistance. This tends to make local site experience with good record keeping the most reliable materials selection guide.

Other materials used in the industry to control abrasive wear include tungsten carbide cermets, ceramics, and hard chromium. Cermets and ceramics are used in either the bulk form (e. g., as inserts, tiles and plates) or as coatings (e. g., as weld overlays or thermal sprays). Hard chromium is used as an electroplated coating. Unlike the stainless steels and cast irons, which are typically used for large items of process equipment, these materials are often found in niche applications where the need to manage particularly demanding conditions justifies their higher cost. Some larger-scale applications of these materials also exist, such as thermal sprayed tungsten carbide cermet coatings for thermorolls on supercalenders, and hard chromium coatings on screen baskets.

References

[1] Tylczak J H. "Abrasive Wear" in Friction, Lubrication and Wear Technology. ASM Handbook, Volume 18, ASM Intl., Materials Park, Ohio 1992: 184-190.

[2] Zum Gahr K H. "Grooving Wear" Chap. 5 in Microstructure and Wear of Materials, Tribology Series. No. 10, Elsevier, New York, 1987.

[3] Budinski K G. Surface Engineering for Wear Resistance. Prentice Hall, Englewood Cliffs, NJ, 1988.

[4] Peterson M B, Winer W O. Eds. Wear Control Handbook, American Soc. of Mechanical Engineers (ASME). New York, 1980.

[5] Borik F. Wear Control Handbook. ASME, New York, 1980: 327-342.

[6] Crook P. Practical guide to wear for corrosion resistance. Mat. Perf. 1991, 30(2): 64-66.

[7] Magee J H. "Wear of Stainless Steels" in Friction, Lubrication and Wear Technology. ASM Handbook, Volume.

[8] ASM Intl. Materials Park, Ohio 1992: 712.

[9] Jones M, Llewellyn R J. Erosion-corrosion assessment of materials for use in the resources industry. Wear, 2009, 267: 2003-2009.

[10] Magee J H. "Wear of Stainless Steels" in Friction, Lubrication and Wear Technology. ASM Handbook, Volume 18, ASM Intl., Materials Park, Ohio 1992: 715.

[11] Thompson C B, Thomson J M, Ko P L. TAPPI, 1994, 77(9): 85-94.

(Source: Chris Thompson. *Stainless Steel and Specialty Alloys for Pulp, Paper and Biomass Conversion*. FP Innovations)

Words and Expressions

martensite *n.* [材] 马氏体，马丁散铁
protrusion *n.* 突出，突出物
abrasion *n.* 磨损，磨耗，擦伤
erosion *n.* 侵蚀，腐蚀，削弱
gouging *n.* 刨削槽，熔刮，凿削
angularity *n.* 倾斜度，有角
humidity *n.* 潮湿，湿气，湿度
overlay *n.* 覆盖物，镶边，研磨料
supercalender *n.* 超级压光机
refiner *n.* 精炼机，精磨机
spray *n.* 喷剂，喷雾器；*v.* 喷涂

13.3 问题与讨论
13.3 Questions and Discussions

(1) What is the difference between grinding and polishing?
(2) Introduce the steps of automatic grinding.
(3) What circumstances use napless cloth and when use napped cloth?
(4) Describe relief polishing, vibratory polishing, and attack polishing.
(5) What circumstances will use manual grinding?

第四部分　热处理

Part 4　Heat Treatment

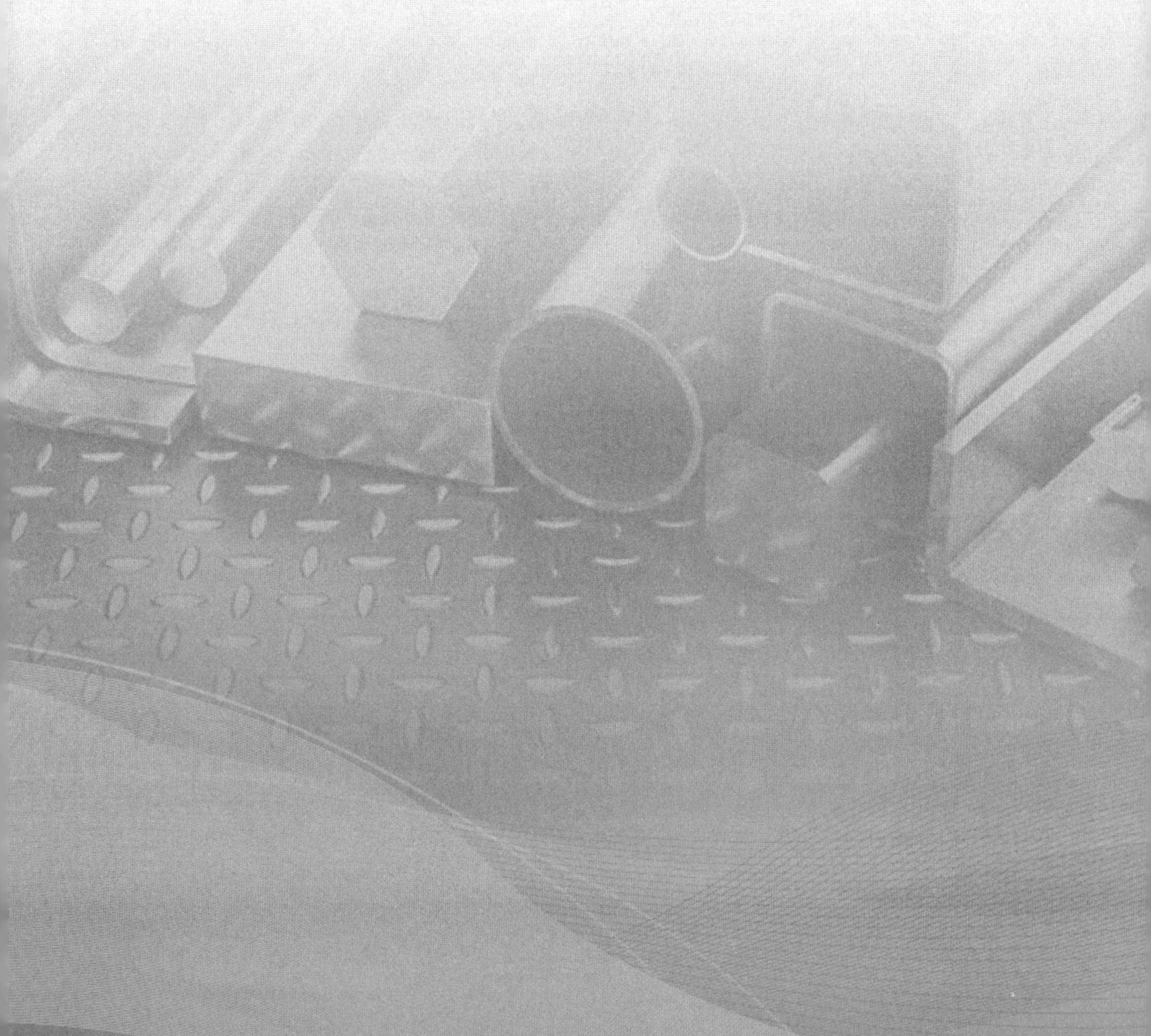

单元 14　金属热处理
Unit 14　Heat Treatment of Metals

扫码查看
讲课视频

14.1　教学内容
14.1　Teaching Materials

1. Introduction

Heat treatment is essentially a process of heating and cooling a material to achieve a desired set of physical and mechanical properties. The properties of materials are dependent upon their structural aspects. In general, heat treatment of metals and alloys concerns the change in microstructure. This in turn alters the mechanical properties of the metals and alloys. During the course of heat treatment, the metals and alloys may undergo change in crystal structure (polymorphic transformation), change in chemical composition and change in the degree of order. Also it will be seen that without any of the above three changes, heat treatment may enable to obtain desired properties of materials by way of changing its substructure. In a nutshell, heat treatment of metals and alloys envisages:

(1) Polymorphic change with or without change in local composition within a system. The average composition however remains the same. This type of change is very common and Fe-C system is a classic example.

(2) Overall change in chemical composition at the surface of a component. This type of change is realized in surface hardening of metal sandalloys.

(3) Change in local composition at a very fine scale of magnitude. Age hardening in alloy exemplifies such incidence.

(4) Dissolution of unwanted second phase particles, and formation of clean single phase alloy.

(5) Grain refinement of a pure metal or a single phase alloy which is previously subjected to mechanical working at a low temperature.

It is to be understood that changes stated above are brought about by previously chosen thermal treatment with an eye to achieving a predefined set of properties. Thus the purposes of heat treatment of metals and alloys are:

(1) To improve strength, hardness, and wear resistance properties.

(2) To increase toughness and ductility.

(3) To obtain fine grain size.

(4) To remove internal stresses induced by differential deformation by cold working, nonuniform cooling from high temperature during casting and welding.

(5) To improve machineability.

(6) To improve cutting properties of tool steels.

(7) To improve surface properties of materials especially in respect of wear and corrosion resistance.

(8) To improve physical properties viz. Electrical and magnetic properties.

(9) To induce other unique material functionalities, some examples are transformation induced plasticity, metamagnetic shape memory effect, etc.

2. Stages of Heat Treatment

The heat treatment as a process is accomplished in the following three stages:

(1) Heating to a high temperature to obtain homogeneous single phase.

(2) Holding at the high temperature for preselected time period. (This process is generally termed "Soaking").

(3) Cooling the object under treatment to room temperature or below.

Again, it may be required to repeat the same three stages with different process parameters. It is necessary to know, how to select the heating temperature. During cooling operation knowledge about the cooling rate to use is also required. If the cooling rate is very slow, equilibrium microstructure in accordance with the prediction of phase diagram of the concerned system is possible to achieve.

3. Common Heat Treating Processes

- **Annealing**

Annealing is the process of heating the object to single phase field followed by equilibrium cooling to aim at achieving equilibrium microstructures. Depending upon the purpose, various annealing types are carried out at different temperature, including full annealing, isothermal annealing, incomplete annealing, spheroidizing annealing, diffusion annealing, recrystallization annealing, subcritical annealing.

- **Normalizing**

Normalizing is the heat treatment process, which consists of heating the object to the austenitization temperature, holding it to produce homogeneous austenite and then allowing to be cooled in still air. Normalizing tends to imply a normal cooling of austenite to produce stronger steel in much shorter time than what can be obtained by full annealing. Due to the fact that normalizing refines the microstructure of the steel thereby improving its toughness, it is quite often employed as the final heat treatment.

- **Hardening**

Hardening is the heat treatment process by which the hardness and hence the strength of a material is increased. Different types of hardening processes are employed in different situations. There are cases of age hardening where the material is heated to get a homogeneous single phase solid solution which is cooled very fast from the high temperature and in the process a supersaturated solid solution is obtained.

- **Quenching**

As stated earlier in case of the Fe-Fe_3C system the supersaturated solid solution is the strongest phase but is extremely brittle. Heating it to some elevated temperature below A_1 temperature decomposes the supersaturated solid solution leading to ductilization of the alloy. By definition the ideal quenching medium will take away the heat from the surface at the same rate as it flows to the surface from inside the bar which is being quenched.

- **Tempering**

As stated earlier the hardened steels with full martensite in microstructure cannot be employed for useful purposes because martensite is extremely brittle; martensite formed by quenching produces high internal stresses in the hardened steel. The hardened steel need be heated to higher temperature to reduce its brittleness; this process is called "Tempering". Tempering is generally carried out by heating the hardened steel to temperatures below A_1; it is held there for some preselected period of time and then followed by cooling to room temperature.

(Source: M K Banerjee. *Fundamentals of Heat Treating Metals and Alloys.* Comprehensive Materials Finishing, 2017)

Words and Expressions

casting *n.* 铸造
machinability *n.* 机械加工性
plasticity *n.* 可塑性
functionality *n.* 功能，[数] 函数性
metamagnetic *a.* 变磁性
homogeneous *a.* 同种类，同性质的，均一的
equilibrium *n.* 平衡；均衡
annealing *n.* 给（金属、玻璃）退火
normalizing *n.* [冶] 正火，正常化
hardening *n.* 硬化，淬火
quenching *n.* 淬火
tempering *n.* 回火，碳钢回火
supersaturate *v.* 过饱和，超饱和
precipitation *n.* 沉淀
austenitization *n.* 奥氏体化
ductilization *n.* 韧塑化，延展性
decomposition *n.* 腐烂，分解
polymorphic transformation [冶] 多晶转变
age hardening 老化变硬，[材] 时效硬化
single phase 单相
grain refinement 晶粒细化
tool steel 工具钢
shape memory effect 形状记忆效应
internal stress 内部应力
spheroidizing annealing 球化退火
diffusion annealing 扩散退火，均匀化退火
recrystallization annealing 再结晶退火
subcritical annealing 不完全（亚临界）退火

Phrases

in a nutshell 简要地说，概括地说
in turn 反过来
bring about 引起，导致，发生
be accomplished in 完成
in accordance with 按照，依照，与…一致
carry out 执行，施行，贯彻
be employed in 任职于，从事于，受雇于

take away　　带走，拿走，取走
in respect of　　关于，涉及

Notes

(1) Heat treatment is essentially a process of heating and cooling a material to achieve a desired set of physical and mechanical properties. The properties of materials are dependent upon their structural aspects.

essentially 译为“本质上，根本上，基本上”；mechanical property 译为“力学性能，机械性能”。

参考译文：热处理本质上是一个加热和冷却材料以达到所需的物理和机械性能的过程。材料的性能取决于它们的结构。

(2) If the cooling rate is very slow, equilibrium microstructure in accordance with the prediction of phase diagram of the concerned system is possible to achieve.

if 引导条件状语从句；in accordance with 译为“依照；与……一致”。

参考译文：如果冷却速度非常慢，则可以实现与相图预测相一致的平衡组织。

(3) Annealing is the process of heating the object to single phase field followed by equilibrium cooling to aim at achieving equilibrium microstructures.

followed by 译为“然后，随后，紧跟着”；aim at 译为“针对，瞄准，目的在于”。

参考译文：退火是将物体加热到单相，然后进行平衡冷却以达到微结构平衡的过程。

(4) Normalizing is the heat treatment process, which consists of heating the object to the austenitization temperature, holding it to produce homogeneous austenite and then allowing to be cooled in still air.

normalizing 译为“正火”，它是冶金中最重要的热处理工艺之一；which 引导定语从句。

参考译文：正火是热处理过程，它是将物体加热到奥氏体温度下保温，使其产生均匀的奥氏体，然后在空气中冷却。

(5) Tempering is generally carried out by heating the hardened steel to temperatures below A_1; it is held there for some preselected period of time and then followed by cooling to room temperature.

tempering 译为“回火”，它是金属热处理的另一重要工艺；carry out 译为“执行，实行，进行”。

参考译文：回火通常是通过将淬火钢加热到 A_1 以下的温度来进行的；它先保持一段预定时间，然后再冷却到室温。

14.2 阅读材料

14.2 Reading Materials

What is Normalizing?

It is important that the material used for any project possesses the correct mechanical properties for the specific application. Heat treatment processes are often used to alter the mechanical properties of a metal, with one of the more common heat treatment processes being Normalizing.

1. What is Normalizing?

Normalizing is a heat treatment process that is used to make a metal more ductile and tough after it has been subjected to thermal or mechanical hardening processes. Normalizing involves heating a material to an elevated temperature and then allowing it to cool back to room temperature by exposing it to room temperature air after it is heated. Normalizing is often performed because another process has intentionally or unintentionally decreased ductility and increased hardness. Normalizing is used because it causes microstructures to reform into more ductile structures. This is important because it makes the metal more formable, more machinable, and reduces residual stresses in the material that could lead to unexpected failure.

2. What is the Difference Between Annealing and Normalizing?

Normalizing is very similar to annealing as both involve heating a metal to or above its recrystallization temperature and allowing it to cool slowly in order to create a microstructure that is relatively ductile. The main difference between annealing and normalizing is that annealing allows the material to cool at a controlled rate in a furnace. Normalizing allows the material to cool by placing it in a room temperature environment and exposing it to the air in that environment.

This difference means normalizing has a faster cooler rate than annealing. The faster cooler rate can cause a material to have slightly less ductility and slightly higher hardness value than if the material had been annealed. Normalizing is also generally less expensive than annealing because it does not require additional furnace time during the cool down process.

3. The Normalizing Process

There are three main stages to a normalizing process.

(1) Recovery Stage. During the recovery stage, a furnace or other type of heating device is used to raise the material to a temperature where its internal stresses are relieved.

(2) Recrystallization Stage. During the recrystallization stage, the material is heated above its recrystallization temperature, but below its melting temperature. This causes new grains without preexisting stresses to form.

(3) Grain Growth Stage. During the grain growth, the new grains fully develop. This growth is controlled by allowing the material to cool to room temperature via contact with air. The result of completing these three stages is a material with more ductility and reduced hardness. Subsequent operations that can further alter mechanical properties are sometimes carried out after the normalizing process.

4. What Metals Can Be Normalized?

To be normalized, a metal needs to be receptive to normalizing, meaning its microstructure can be altered by heat treatment. Many types of alloys can be normalized, including: iron based alloys (tool steel, carbon steel, stainless steel, and cast iron); Nickel-based alloys; Copper; Brass; Aluminum.

5. Common Applications for Normalizing

The low cost of the normalizing process makes it one of the most extensively used industrial process when compared to annealing. The furnace is available for the next batch as soon as heating and holding periods are over. Normalizing is used to:

(1) Improve the grain size refinement and machinability of cast structures of castings.

(2) Recover the original mechanical properties of forged or cold worked steel.

(3) Ease the forging operations for high carbon steel.

(4) Stress relieve of castings.

Normalizing is used in many different industries for many different materials. Examples include:

(1) Ferritic stainless steel stampings in the automotive industry may be normalized following the work hardening that occurs during their forming process.

(2) Nickel-based alloys in the nuclear industry may be normalized following the thermal microstructure alteration that occurs following welding.

(3) Carbon steel may be normalized after it is cold-rolled to reduce the brittleness caused by work hardening.

6. Normalizing Heat Treatment & Process

Normalizing heat treatment is a process applied to ferrous materials. The objective of the normalizing heat treatment is to enhance the mechanical properties of the material by refining the microstructure. The ferrous metal is heated to the austenite phase, above the transformation range, and is subsequently cooled in still air at room temperature.

The metal is heated in a furnace for normalizing heat treatment process (Figure 1). The temperature of the furnace is kept between 750~980℃ (1320~1796℉) also known as "holding temperature", depending upon the carbon content in the material. The material is kept at the temperature above austenite temperature for 1~2 hours, until all the ferrite converts into austenite, and then cooled to room temperature in still air or Nitrogen, if run in the vacuum furnace at less than 1 bar pressure. This process produces fine pearlite structure which is more uniform. Pearlite is a layered structure of two phases i. e. cementite (iron carbide) and α-ferrite. Normalizing heat treatment produces a more uniform carbide size which aids further heat treatment operations and results in a more consistent final product. Normalized steel has greater strength and hardness than

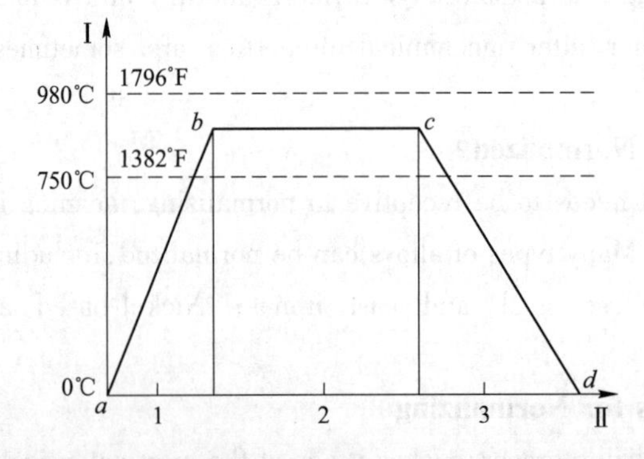

Figure 1　Normalizing heat treatment process

annealed steel, and the process is more economical due to cooling directly with air. The process of normalizing is explained in following. The metal is heated from temperature "*a*" to "*b*" and kept in this condition for some time. It is then cooled to ambient temperature "*d*" in still air.

7. Microstructure in Normalizing

The thickness of carbon steel can have a significant effect on the cooling rate and thus the resulting microstructure. The thicker pieces cool down slower and become more ductile after normalizing than thinner pieces. After normalizing the portions of steel containing 0.80% of carbon are pearlite while the areas having low carbon are ferrites. The redistribution of carbon atoms takes place between ferrite (0.022 wt%) and cementite (6.7 wt%) by the process of atomic diffusion (Figure 2). The amount of pearlite is more than that in annealed steel with same carbon content. This is because of shifting of the eutectoid composition to lower value and formation of cementite. The fine-grained pearlite microstructure is tougher than coarse-grained ones. Normalizing reduces the internal stresses of the carbon steel. It also improves microstructural homogeneity, enhances thermal stability and response to heat treatment.

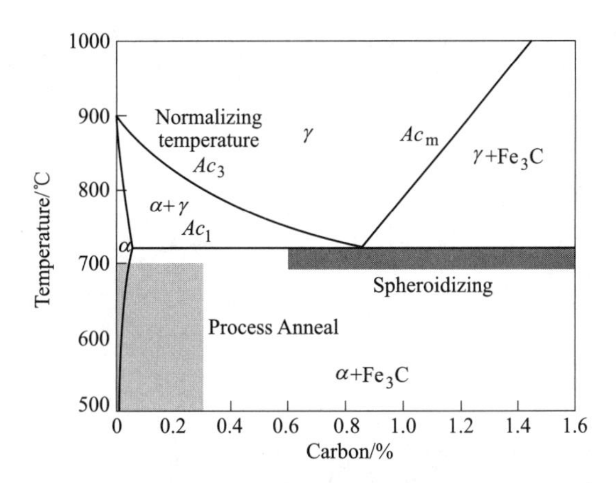

Figure 2　Iron-carbon phase diagram

8. Normalizing Equipment

The equipment in use for normalizing comes in both batch and continuous operations. Bell furnace offers an economical method of heat treatment and different bell lifting mechanisms. Continuous furnaces heat treats the metal in the continuous fashion. The conveyor runs at constant speed, and the product is carried to desired conditions after heat treatment.

(Source: *What is Normalising?* Metalsupermarket Website)

Words and Expressions

brittleness　*n.* 脆性

residual stress　残余应力

nuclear industry　[核] 核工业，原子能工业

ambient temperature　环境温度,室温

14.3 问题与讨论

14.3 Questions and Discussions

(1) What is heat treatment and what will happen during heat treatment?

(2) What is the purposes of heat treatment of metals and alloys?

(3) The basic steps of heat treatment.

(4) Common heat treating processes.

(5) The principle of normalizing and quenching.

单元 15　表面淬火
Unit 15　Surface Hardening

扫码查看
讲课视频

15.1　教学内容
15.1　Teaching Materials

Surface hardening, a process that includes a wide variety of techniques (Table 1), is used to improve the wear resistance of parts without affecting the more soft, tough interior of the part. This combination of hard surface and resistance to breakage upon impact is useful in parts such as a cam or ring gear, bearings or shafts, turbine applications, and automotive components. Further, the surface hardening of steel can have an advantage over through hardening because less expensive low-carbon and medium carbon steels can be surface hardened with minimal problems of distortion and cracking associated with the through hardening of thick sections. There are two distinctly different approaches to the various methods for surface hardening (Table 1).

Table 1　Engineering methods for surface hardening of steels

Layer additions	Substrate treatment
1. Hardfacing:	1. Diffusion methods:
(1) Fusion hardfacing	(1) Carburizing
(2) Thermal spray	(2) Nitriding
2. Coatings.	(3) Carbonitriding
(1) Electrochemical plating	(4) Nitrocarburizing
(2) Chemical vapor deposition (electroless plating)	(5) Boriding
(3) Thin films (physical vapor deposition, sputtering, ion plating)	(6) Titanium-carbon diffusion
(4) Ion mixing	(7) Toyota diffusion process
	2. Selective hardening methods:
	(1) Flame hardening
	(2) Induction hardening
	(3) Laser hardening
	(4) Electron beam hardening
	(5) Ion implantation
	(6) Selective carburizing and nitriding
	(7) Use of arc lamps

The first group of surface-hardening methods includes the use of thin films, coatings, or weld overlays (hardfacings). Films, coatings, and overlays generally become less cost-effective as production quantities increase, especially when the entire surface of workpieces must be hardened. Some overlays can impart corrosion-resistant properties. This introductory article on surface hardening focuses exclusively on the second group of methods, which is further divided into diffusion methods and selective-hardening methods (Table 1). Surface hardening by diffusion involves the chemical modification of a surface. The basic process used is thermochemical because some heat is needed to enhance the diffusion of hardening elements into the surface and subsurface regions of a part. Diffusion methods modify the chemical composition of the surface with hardening species such as carbon, nitrogen, or boron. Diffusion methods may allow effective hardening of the entire surface of a part and are generally used when a large number of parts are to be surface hardened. The depth of diffusion exhibits a time-temperature dependence such that:

$$\text{Case depth} = K\sqrt{\text{Time}} \tag{1}$$

where the diffusivity constant, K, depends on temperature, the chemical composition of the steel, and the concentration gradient of a given hardening element. In contrast, selective surface-hardening methods allow localized hardening. Selective hardening generally involves transformation hardening (from heating and quenching), but some selective-hardening methods (selective nitriding, ion implantation, and ion beam mixing) are based solely on compositional modification.

1. Carburizing and Carbonitriding

Carburizing is the addition of carbon to the surface of low-carbon steels at temperatures (generally between 850 and 980℃) at which austenite, with its high solubility for carbon, is the stable crystal structure. With grades of steel engineered to resist grain coarsening at high temperatures and properly designed furnaces such as vacuum furnaces, carburizing above 980℃ is practical to dramatically reduce carburizing time. Hardening is accomplished when the high-carbon surface layer is quenched to form martensitic case with good wear and fatigue resistance superimposed on a tough, low-carbon steel core. Of the various diffusion methods, gas carburization is the most widely used, followed by gas nitriding and carbonitriding.

2. Nitriding and Nitrocarburizing

Nitriding is a surface-hardening heat treatment that introduces nitrogen into the surface of steel at a temperature range (500 to 550℃), while it is in the ferritic condition. Because nitriding does not involve heating into the austenite phase with quenching to form martensite, nitride components exhibit minimum distortion and excellent dimensional control. Nitriding has the additional advantage of improving corrosion resistance in salt spray tests. Higher-alloying elements usually retard the N_2 diffusion rate, which slows the case depth development. Thus, nitriding requires longer cycle times to achieve a given case depth than that required for carburizing.

3. Applied Energy Methods

Surface hardening of steel can be achieved by localized heating and quenching, without any chemical modification of the surface. The more common methods currently used to harden the surface of steels include flame and induction hardening. However, each of these methods has shortcomings that can prevent its use in some applications.

(1) Flame hardening consists of austenitizing the surface of a steel by heating with an

oxyacetylene or oxyhydrogen torch and immediately quenching with water or water-based polymer. The result is a hard surface layer of martensite over a softer interior core with a ferrite-pearlite structure. There is no change in composition, and therefore, the flame-hardened steel must have adequate carbon content for the desired surface hardness.

(2) Induction heating is an extremely versatile heating method that can perform uniform surface hardening, localized surface hardening, through hardening, and tempering of hardened pieces. Heating is accomplished by placing a steel ferrous part in the magnetic field generated by high-frequency alternating current passing through an inductor, usually a water-cooled copper coil.

(3) Laser surface heat treatment is widely used to harden localized areas of steel and cast iron machine components. This process is sometimes referred to as laser transformation hardening to differentiate it from laser surface melting phenomena. There is no chemistry change produced by laser transformation hardening, and the process, like induction and flame hardening, provides an effective technique to harden ferrous materials selectively.

(4) Electron beam (EB) hardening, like laser treatment, is used to harden the surfaces of steels. The EB heat treating process uses a concentrated beam of high-velocity electrons as an energy source to heat selected surface areas of ferrous parts.

(Source: Michael J Schneider, et al. *Introduction to Surface Hardening of Steels*. ASM International Website)

Words and Expressions

cam *n.* 凸轮
gear *n.* 齿轮，器械，装置
bearing *n.* 轴承，支轴
shaft *n.* 轴，杆
turbine *n.* 涡轮，涡轮机
hardfacing *n.* 耐磨堆焊，硬面堆焊
boriding *n.* [材] 渗硼，硼化
nitriding *n.* 渗氮，渗氮法
carbonitriding *n.* 碳氮共渗
nitrocarburizing *n.* 渗碳氮化，碳氮共渗
diffusivity *n.* 扩散率，扩散性
constant *n.* 常数
gradient *n.* 梯度，倾斜度，坡度
exponentially *ad.* 以指数方式
oxyacetylene *n.* 氧乙炔
versatile *a.* 多用途的，多功能的，万能的
inductor *n.* 感应器
surface hardening 表面硬化，表面淬化
through hardening 整体硬化，穿透淬火
electrochemical plating 电化学镀层
chemical vapor deposition 化学气相沉积
selective hardening 局部淬火，选择硬化
ion implantation 离子注入
grain coarsening 晶粒粗化
induction hardening 感应硬化，感应淬火
case depth 渗层厚度
laser transformation hardening 激光相变硬化

Phrases

combination of 结合了，综合了
divide into 分成
be practical to ……是实用的

Notes

(1) The surface hardening of steel can have an advantage over through hardening because less expensive low-carbon and medium carbon steels can be surface hardened with minimal problems of distortion and cracking associated with the through hardening of thick sections.

surface hardening（表面硬化）和 through hardening（穿透硬化）两者是钢硬化的不同处理工艺；have advantages over 译为“优于”。

参考译文：此外，钢的表面硬化比穿透硬化更具有优势，因为较便宜的低碳和中碳钢表面硬化时与厚截面的穿透硬化相比，其变形和裂纹问题更小。

(2) In contrast, selective surface-hardening methods allow localized hardening. Selective hardening generally involves transformation hardening (from heating and quenching), but some selective-hardening methods (selective nitriding, ion implantation, and ion beam mixing) are based solely on compositional modification.

in contrast 译为“相比而言”；localized hardening 译为“局部淬火，局部硬化”；transformation hardening 译为“相变硬化”。

参考译文：相对而言，选择性表面硬化允许局部硬化。选择性表面硬化通常涉及相变硬化（来自加热和淬火），但是某些选择性硬化方法（选择性氮化，离子注入和离子束混合）仅基于成分修饰。

(3) Carburizing is the addition of carbon to the surface of low-carbon steels at temperatures (generally between 850 and 980℃) at which austenite, with its high solubility for carbon, is the stable crystal structure.

low-carbon steels 译为“低碳钢”；at which 引导定语从句，修饰“temperature”。

参考译文：渗碳是在一定温度下（通常在850~980℃）向低碳钢的表面添加碳，在该温度下，对碳具有高溶解度的奥氏体是稳定的晶体结构。

(4) Hardening is accomplished when the high-carbon surface layer is quenched to form martensitic case with good wear and fatigue resistance superimposed on a tough, low-carbon steel core.

superimposed on 译为“在……上叠加”；when 引导宾语从句。

参考译文：当高碳表面层淬火形成耐磨性和抗疲劳性良好的马氏体外壳叠加在一个坚韧的低碳钢芯上时，淬火就完成了。

(5) Nitriding has the additional advantage of improving corrosion resistance in salt spray tests. Higher-alloying elements usually retard the N_2 diffusion rate, which slows the case depth development. Thus, nitriding requires longer cycle times to achieve a given case depth than that required for carburizing.

nitriding(渗氮) 和 carburizing(渗碳) 是两种常用的表面处理工艺；case depth 指的是渗氮（碳）深度或硬化层深度。

参考译文：渗氮还有另一个优点，就是可以提高盐雾试验中的耐腐蚀性。高合金元素会阻碍 N_2 扩散速度，从而减慢渗层深度的发展。因此，与渗碳相比，渗氮需要更长的循环时间才能达到给定的渗碳深度。

(6) There is no chemistry change produced by laser transformation hardening, and the process, like induction and flame hardening, provides an effective technique to harden ferrous materials selectively.

laser transformation hardening(激光相变硬化)、induction hardening(感应硬化) 和 flame hardening(火焰淬火) 是其他一些表面处理工艺，这里 hardening 也是"淬火"之意。

参考译文：激光相变硬化不产生化学变化，就像感应淬火和火焰淬火等工艺一样，提供了一种黑色金属材料硬化的有效技术。

15.2 阅读材料
15.2 Reading Materials

Surface Coating Techniques

The advantage of coating technology, in general, is that it marries two dissimilar materials to improve, in a synergistic way, the performance of the whole. Usually mechanical strength and fracture toughness are being provided by the substrate and the coating provides protection against environmental degradation processes including wear, corrosion, erosion, and biological and thermal attack.

Surface coating technologies have the following advantages.

(1) Technical advantages. Creation of new materials (composites) with synergistic property enhancement, or completely new functional properties, for example electronic conductivity, piezo-or ferroelectric properties etc.

(2) Economic advantages. Expensive bulk materials such as stainless steel or superalloys can be replaced by relatively thin overlays of a different material. These savings are enhanced by longer lifetime of equipment and reduced downtime and shortened maintenance cycles.

(3) Attitudinal advantages. Materials engineers trained in metals handling need not be afraid to deal with new materials with unfamiliar properties, specifications and performance such as ceramics or polymeric composites. The ceramic coating just becomes a part of a familiar metal materials technology.

Coatings are, of course, not a new invention. For times immemorial, wood and metal have been painted with or ganic or inorganic pigments to improve their esthetic appearance and their environmental stability. Corrosion-, wear- and abrasion resistance of the substrate materials were significantly improved by the paint coatings. These organic paint coatings, however, did not endure high temperatures and did not adhere well. The performance of traditional coatings has been improved by the use of chemically-cured paints in which components are mixed prior to application and polymerized by chemical interaction, i. e. cross-linking. Epoxy resins, polyurethanes and various polyester finishes show considerable resistance to alkalis and acids, and also to a wide range of oils, greases and solvents.

Traditional enamels are glass-based coatings of inorganic composition applied in one or more layers to protect steel, cast iron or aluminum surfaces from corrosion. This technology has been extended today to manufacture thick film electronics in which metals or metal oxides are added to a fusible glass base to generate a range of thick film conductors, capacitors and other electronic components.

Chemical coatings are frequently applied by electroplating of metals such as copper and nickel. Nickel, for example, forms a highly adherent film for wear applications by an electroless plating process, and thus is applied to manufacture aerospace composites. A related technology is the anodizing of aluminum by electrochemically induced growth of aluminium oxide in a bath of sulphuric or phosphoric acids.

Spray pyrolysis involves chemical reactions at the surface of a heated substrate. Increasingly transparent conducting coatings of tin oxide or indium tin oxide are used to coat glass windows for static control, radio frequency shielding and environmental temperature control.

Sol-gel coatings based on the pyrolysis of organometallic precursors such as metal alkoxides are used today. The process was originally developed for aluminum and zirconium oxide but is now extending to a wide range of glasses including silicates and phosphates, and has recently been applied to complex ferroelectrics such as PZT (lead zirconate titanate) and PLZT (lead lanthanum zirconium titanate). Sol-gel coatings enjoy a high compositional flexibility and ease of preparation at generally ambient temperature but because of the frequently expensive precursor materials their application is limited to high-value added devices, in particular in electronics.

Thin coatings produced by Chemical Vapor Deposition (CVD) are widely employed in the semiconductor industry for large band-gap materials such as gallium arsenide, indium phosphide and other compound semiconductor materials. The technology uses vapor phase transport to grow epitaxial and highly structured thin films including insulating oxide films on single crystal silicon substrates. A related technology is the growth of thin crystalline diamond films by decomposition of methane or other hydrocarbons in a hydrogen ($>95\%$)-argon ($<5\%$) plasma. Much activity is currently devoted in Japan and the USA to the improvement of the thickness and the crystallographic perfection of diamond thin films. Major potential industrial applications of such films can be found for protective coatings on compact discs, optical lenses, in particular such carried by low-earth orbit space craft, and substrates for ULSI (ultra large scale integration) devices.

Physical Vapor Deposition (PVD) technologies using evaporation, sputtering, laser ablation, and ion bombardment are a mainstay of present-day surface engineering technology.

(1) Evaporation is the most simple vacuum technique. The materials to be deposited on a substrate are melted and vaporized either on a resistively heated tungsten or molybdenum boat, or by an electron beam. The method is suitable for many metals, some alloys, and compounds with a high thermal stability such as silica, yttria and calcium fluoride. While films deposited by evaporation are inferior to other vacuum techniques in terms of adhesion to the substrate, the excellent process control generating optical films with well-defined thicknesses and indices of refraction has made this technique popular.

(2) Sputtering methods deposit material by causing atoms to separate from a target by bombardment with highly energetic ions from a gas plasma or a separately exited ion beam, and depositing them on a substrate. Magnetron sputtering uses confinement of the exiting plasma by a strong magnetic field that results in high deposition rates and good reproducibility. Large area sources are used to coat plastic foils and ceramic substrates for packaging materials in the food industry, as well as for metallizing plastic ornaments and automotive components such as bumpers, for architectural glass, and for multilayer holographic coatings on identification and credit cards to prevent forgeries.

(3) Laser ablation is a modern technology that has been developed in particular for high temperature superconducting ceramics. A focused laser beam is used to vaporize the target material. Since this material comes from a highly localized region the ion flux reproduces the target composition faithfully. Even though the cost of the lasers is still high and the deposition rates are quite low, future developments may lead to much wider application of this technique if reproducible process control can be achieved.

The PVD methods mentioned so far all rely on a coating deposited on an existing substrate surface with given composition. However, ion implantation modifies the properties of the substrate itself. A beam of high energy ions can be created in an accelerator and brought in contact with a substrate surface. Thus corrosion and wear performance can be improved dramatically. For example, implantation of 18% chromium into a steel results in an in-situ stainless steel with high corrosion resistance, the use of boron and phosphorus produces a glassy surface layer inhibiting pitting corrosion, and the implantation of titanium and subsequent carbon ions creates hard-phase titanium carbide precipitates. Likewise, implantation of nitrogen produces order of magnitude improvement of the wear resistance of steel, vanadium alloys, and even ceramics.

Modern high performance machinery, subject to extremes of temperature and mechanical stress, needs surface protection against high temperature corrosive media, and mechanical wear and tear. For such coatings a highly versatile, low cost technique must be applied that can be performed with a minimum of equipment investment and does not require sophisticated training procedures for the operator. Such a technique has been found in thermal spraying. It uses partially or complete melting of a wire, rod or powder as it passes through a high temperature regime generated electrically by a gas plasma or by a combustion gas flame. The molten droplets impinge on the substrate and form the coating layer by layer. This technique is being used widely to repair and resurface metallic parts and also, in recent years, to build up wear-, chemical- and thermal barrier coatings based on alumina or zirconia, in particular for applications in the aerospace and automotive industries. High temperature-erosion protection of boiler tubes and fire chambers of coal-fired power plants, corrosion protection of special concrete parts, and of bioceramic coatings for orthopedic and dental prosthetic implants are just a short list of ever-increasing fields of service of thermally sprayed coatings.

For all these reasons, advanced materials, i. e. ceramics or polymer coatings, become more and more popular among materials engineers. The equipment, ease of application of coatings to complex surfaces, and the availability of tailored materials makes novel surface coatings increasingly attractive. Thus, it has been said that coatings technology will be the materials technology of the 21st century.

(Source: Robert B Heimann. *Plasma-Spray Coating*. VCH Verlagsgesellschaft mbH, 1996)

Words and Expressions

specification　*n.* 规格，规范，说明书
enamel　*n.* 搪瓷，珐琅，瓷釉
capacitor　*n.* ［电］电容器
sulphuric　*a.* 含硫黄的，含硫的
phosphoric　*a.* 磷的，含磷的
pyrolysis　*n.* ［化学］热解，高温分解
sol-gel　*n.* 溶胶-凝胶
precursor　*n.* 前驱体，前质
alkoxide　*n.* ［化］醇盐，酚盐
lanthanum　*n*, ［化］镧（La）
titanate　*n.* ［化］钛酸盐
epitaxial　*a.* 外延的，取向附生的
methane　*n.* 甲烷，沼气
hydrocarbon　*n.* ［化］碳氢化合物
yttria　*n.* ［化］氧化钇
ornament　*n.* 装饰品，点缀品
bioceramic　*n.* 生物陶瓷
combustion　*n.* 氧化，燃烧
prosthetic　*n.* 假肢，假体
ever-increasing　*a.* 不断增长的
thermal attack　热侵蚀
fracture toughness　断裂韧性
electronic conductivity　电导率
inorganic pigment　无机颜料
chemically cured　化学交联，固化
abrasion resistance　抗磨损性
gallium arsenide　砷化镓
laser ablation　激光轰击，激光烧蚀
electron beam　电子束
calcium fluoride　［化］氟化钙
magnetron sputtering　磁控溅射
physical vapor deposition　物理气相沉积
thermal spraying　热喷涂

15.3　问题与讨论

15.3　Questions and Discussions

(1) What is surface hardening and what is it usually used for?
(2) Introduce two different approaehes for surface hardening.
(3) What is the relationship between diffusion depth and time?
(4) What is selective surface-hardening?
(5) What is carburizing and nitriding?
(6) What is the common methods currently used to harden the surface of steels?

单元 16 金属合金铸造
Unit 16 Casting of Metal Alloys

扫码查看
讲课视频

16.1 教学内容
16.1 Teaching Materials

1. Shape and Ingot Castings

Casting is the operation of pouring molten metal into amould and allowing it to solidify. The pouring temperature is usually 50～180℃ above the melting point of the metal alloy. There are two broad classes of casting operations known as shape casting and ingot casting. Shape casting involves pouring the liquid metal into a mould having a shape close to the geometry of the final component. The shape casting is removed from the mould after solidification, and then heat treated and machined into the finished component. Ingot casting, on the other hand, involves pouring the metal into a mould having a simple shape, such as a bar or rod. After cooling, the ingot is strengthened and shaped by working processes such as forging, extrusion or rolling.

2. Solidification of Castings

Solidification of the metal within shape or ingot casting moulds has a strong influence on the microstructure and mechanical properties. The solidification process is complex and does not simply involve the metal changing immediately from liquid to solid when cooled to the freezing temperature. When molten metal is cooled below the equilibrium freezing temperature there is a driving force for solidification and it might be expected that the liquid would spontaneously solidify. However, this does not always occur—pure metals can be cooled well below their equilibrium freezing temperature without solidifying. This phenomenon is known as undercooling or supercooling. The reason for this behaviour is that the transformation from liquid to solid begins with the formation of tiny solid particles or nuclei within the melt. The nuclei develop spontaneously in the liquid as nano-sized particles composed of several dozen atoms arranged in a crystalline structure. The nuclei are freely suspended in the melt and are surrounded by liquid metal. The creation of solid nuclei in this way is called homogenous nucleation. The nuclei often dissolve back into liquid before they grow to a critical size that is thermodynamically stable. It is only when nuclei grow beyond a critical size, which requires cooling of the liquid well below the equilibrium freezing temperature, that the solidification process stabilises and the metal completely solidifies.

In practice, large undercooling does not occur with metal alloys because solidification occurs by a process called heterogeneous nucleation. Most alloys freeze within about 1℃ of their equilibrium melting temperature because the nucleation of ultra-fine solid nuclei occurs at free surfaces, such as the mould walls or solid impurity particles in the melt. The heterogeneous nucleation of solid nuclei at pre-existing surfaces gives them greater stability than the nuclei which develop by homogenous nucleation. Once the nuclei have formed by heterogeneous nucleation they grow by atoms within the liquid attaching to the solid surface. At the equilibrium freezing temperature, a metal contains an extremely large number of nuclei. The sequence of events in the solidification of a metal under the conditions of heterogeneous nucleation is presented in Figure 1.

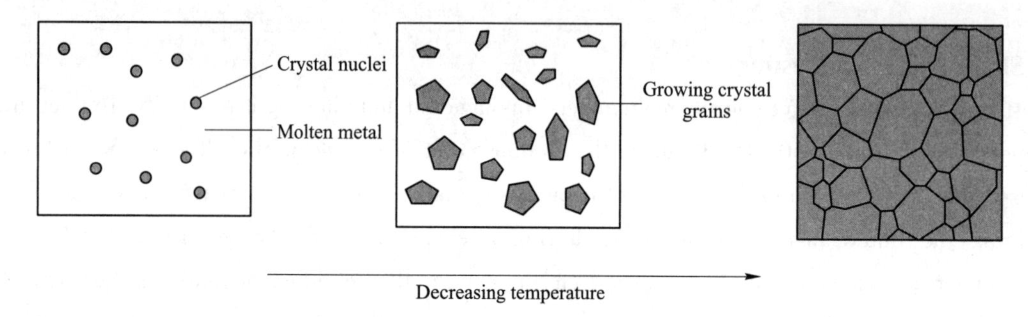

Figure 1　Solidification sequence for a metal

3. Structure of Castings

The solidification of shape and ingot castings often occurs in three separate phases, with each phase developing a characteristic arrangement of grain sizes and shapes. The grain structures through the section of an ingot from the mould wall to the centre are shown in Figure 2. From the surface to core these regions are called the chill zone, columnar zone and central zone, and each has a distinctive grain structure.

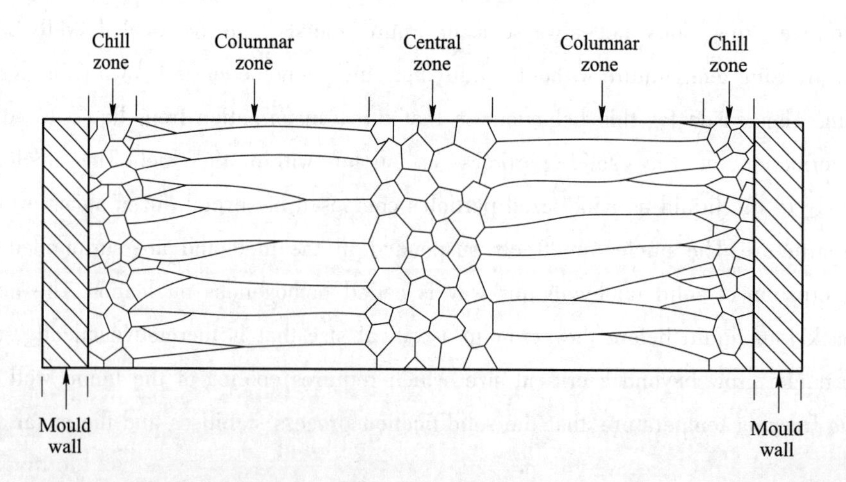

Figure 2　Section through a casting showing the formation of the grain structures during solidification

The chill zone is a thin band of randomly orientated grains at the surface of the casting. This is the first phase to develop during solidification because the mould surface rapidly cools the metal below the melting temperature. Many solid nuclei particles are created by heterogeneous nucleation at the mould wall and grow into the liquid to form the chill zone. When the pouring temperature of the molten metal into the mould is too low, the chill zone, which consists of equiaxed grains, extends through the entire casting and no other zones develop. However, most casting is performed with a high pouring temperature that keeps the metal at the centre above the melting temperature for a long time. Under these conditions, the columnar zone develops from the chill zone.

The columnar zone forms by the solidification of liquid metal into elongated grains. The columnar grains grow perpendicular to themould wall by a solidification process involving dendrites. Secondary and tertiary dendrite arms grow out from the primary stalk, and the branched dendrite structure often appears like a miniature pine tree.

The metal sometimes continues to solidify in a columnar manner until all the liquid has solidified. More often, however, a central zone (also called the equiaxed zone) develops at the core of the casting. The equiaxed zone contains relatively round grains with a random orientation, and they often stop the growth of columnar grains. The amount of the final cast structure that is columnar or equiaxed depends on the cooling rate and alloy composition. A sharp thermal gradient across the solid-liquid interface owing to rapid cooling encourages columnar solidification whereas a low thermal gradient promotes equiaxed solidification.

(Source: Adrian P Mouritz. *Introduction to Aerospace Materials*. Cambridge. Woodhead Publishing, 2012)

Words and Expressions

solidify *v.* 固化，凝固
bar *n.* 圆棒
rod *n.* 方形条杆
supercooling *a.* 过冷的
nuclei *n.* 核心，原子核（nucleus 的复数）
heterogeneous *a.* 不均一的，多相的
nucleation *n.* 形核，成核现象
equiaxed *a.* 各向等大的，等轴的
dendrite *n.* 树突，枝晶
stalk *n.* 茎秆
branched *a.* 分枝的，枝状的
tertiary *a.* 第三的，次要的，[化] 叔的
shape casting 成型铸造
freezing temperature 冰点，凝固温度
driving force 驱动力
homogenous nucleation [晶] 均相成核
heterogeneous nucleation [晶] 异相成核
chill zone 表面等轴晶区，冷硬区
columnar zone 柱状晶区
pouring temperature 浇注温度

Phrases

allow to 允许，使之……
on the other hand 另一方面

in practice　　实际上，事实上
be surrounded by　　被……包围，包裹

Notes

(1) Shape casting involves pouring the liquid metal into a mould having a shape close to the geometry of the final component. The shape casting is removed from the mould after solidification, and then heat treated and machined into the finished component.

shape casting 译为“成型铸造”；pour into 译为“倒入……中”。

参考译文：成型铸造包括将液态金属浇注到形状接近最终部件几何形状的模具中，凝固后从模具中取出成型铸件，然后进行热处理并加工成成品构件。

(2) When molten metal is cooled below the equilibrium freezing temperature there is a driving force for solidification and it might be expected that the liquid would spontaneously solidify.

freezing temperature 译为“冰点，凝固温度”；driving force 译为“驱动力”；be expected 译为“有望……，预期……”；when 引导条件状语从句。

参考译文：当熔融金属冷却到平衡凝固温度以下时，就会产生一个凝固驱动力，预期液体可以自发凝固。

(3) It is only when nuclei grow beyond a critical size, which requires cooling of the liquid well below the equilibrium freezing temperature, that the solidification process stabilises and the metal completely solidifies.

when 引导条件状语从句；which 引导定语从句，修饰布 Critical size。

参考译文：只有当液体冷却到远低于平衡凝固温度，晶核长大超过临界尺寸时，凝固过程才趋于稳定，金属才会完全凝固。

(4) The chill zone is a thin band of randomly orientated grains at the surface of the casting. This is the first phase to develop during solidification because the mould surface rapidly cools the metal below the melting temperature.

chill zone 本意为“冷硬区”，实际就是常说的“表层细晶区”。

参考译文：表层细晶区是铸件表面随机取向的细晶粒带。这是凝固过程中形成的第一个阶段，因为模具表面会将金属快速冷却至低于熔化温度。

(5) The amount of the final cast structure that is columnar or equiaxed depends on the cooling rate and alloy composition. A sharp thermal gradient across the solid-liquid interface owing to rapid cooling encourages columnar solidification whereas a low thermal gradient promotes equiaxed solidification.

owing to 译为“由于，因为”；encourage 此处为“促进”之意，相当于后面的 promote。

参考译文：最终铸态组织呈柱状或等轴状的数量取决于冷却速度和合金成分。快速冷却使固-液界面产生强烈的温度梯度促进柱状凝固，而低的温度梯度促进等轴凝固。

16.2 阅读材料
16.2 Reading Materials

Overview of Metal Injection Moulding

Metal injection moulding(MIM) is a metalworking technology for cost-effectively producing small, complex, precision metal parts in high run volumes that is currently receiving a great deal of attention. The process was first noted in the early 1970s and is generally attributed to the Parmatech Corporation of Petaluma, CA. In May 1987, the Metal Injection Moulding Association (MIMA) was formed, under the auspices of the Metal Powder Industries Federation (MPIF). Currently, eighteen companies are members, including both in-plant and contract fabricators.

Metal injection moulding is a marriage of thermoplastic injection moulding and conventional powder metallurgy technologies. It provides an alternative for the manufacture of small, complex, precision metal parts, which can be produced cost-effectively in high run volumes. Like all processes, MIM has certain advantages and disadvantages. The limits of selected design parameters will be reviewed, along with the general process and MIM's competitive position relative to other metalworking technologies.

1. The MIM Process

The metal injection moulding process begins with very fine metal powders, typically less than 20 microns in diameter. For comparison, compaction grade powders for conventional PM can range up to about 200 microns. While there are serious efforts underway to use coarser powder, material of less than 10 microns is presently preferred by most MIM fabricators. For ferrous compositions, where the alloy will be formed through elemental additions, carbonyl iron powder is commonly used. In order to form a feedstock, the metal powders are mixed in precise proportions with expendable thermoplastic binder and other additives. Those polymeric additions often comprise as much as 40 vol% of the feedstock. A review of the literature indicates that polyethylene or polypropylene are typical ingredients. In practice, feedstocks are multi-component systems, composed of a number of "binder-like" additions.

A flow chart illustrating the steps of the MIM process is shown in Figure 1. The blend of appropriate powders and polymeric binders is granulated, and then is fed into a standard injection moulding machine. Under moderate temperatures and pressures (less than 500℉/260℃ and less than 10000psi/69MPa), the feedstock achieves a toothpaste-like consistency and can be injected into mould cavities to form precision components. After removal from the mould, the parts have excellent green strength and exhibit a crayon-like consistency. The binders and additives used in the moulding step are now removed by processing which has come to be known as debinderizing. Most commonly, this consists of a solvent extraction or a low temperature thermal treatment to breakdown and vaporize the polymers. Binders are designed to come off sequentially so as not to disturb the geometry of the part, leaving just enough binder to hold the metal shape together. Finally, the component is sintered, but at somewhat higher temperatures and longer times than for conventional powder metallurgy.

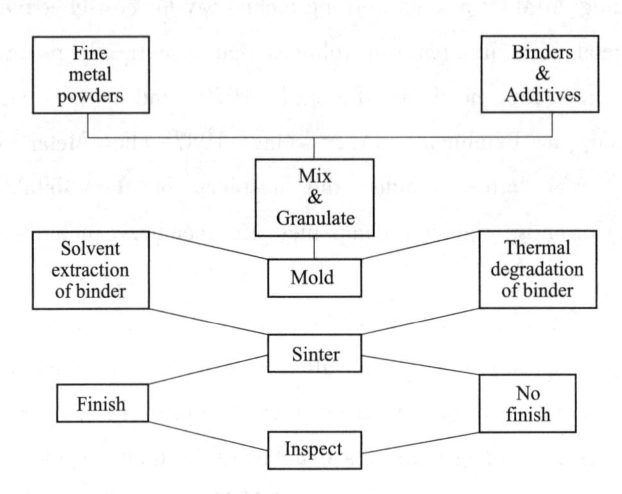

Figure 1　Flow diagram of the MIM process

Initially, parts were sintered in batch furnaces or retorts; today, most fabricators are using continuous, high temperature atmosphere sintering furnaces similar to those used for PM. The very fine powders used for MIM provide a strong thermodynamic driving force and parts undergo shrinkage of as much as 20%. Repeatability is achieved through close control over the feedstock and moulding variables. Once sintered, MIM parts can be given supplementary treatments exactly like wrought products, including heat treatment, plating or machining.

2. Competitive Position

The MIM process is well suited to the high volume production of geometrically complex parts. Based on experience obtained in serving the commercial markets. MIM appears to compete primarily with smaller investment castings and discrete machining. It also competes with conventional PM when machining of the PM part is required to generate additional geometry. Assemblies made by combining a number of small parts can often be moulded as an integral piece by MIM. The process is capable of forming complex configurations, and has a degree of geometric freedom equivalent to that of injection moulding plastics. In addition to part complexity, MIM offers the capability of providing smooth surfaces, close tolerance and thin-walled sections. The limits of these characteristics will be defined in later sections.

3. Materials

With MIM, alloy compositions can be blended to most specific end-use requirements. In fact, almost any metal or ceramic which can be obtained as fine particles and which can be sintered can be injection moulded. However, the compositions which are most readily available from current part suppliers are the iron-nickel alloys and stainless steels. Nickel steels range from iron with 2% nickel for structural components to iron with 50% nickel for magnetic applications. Other magnetic applications are being satisfied with a 3% silicon steel alloy. The most common stainless steel is 316, although other 300 series and some 400 series stainless compositions are available along with 17-4 PH.

Table 1 shows typical properties of selected injection moulded alloys. Many fabricators process material to very low carbon levels in order to achieve reproducibility. It is anticipated that as the industry grows, a wide range of alloys will become available.

Table 1 Typical properties of injection moulded alloys

Property	Composition			
	Fe-2%N	Fe-3%Si	316 Stainless	410 Stainless
Density (g/cc)	7.5	7.8	7.5	7.5
Full density/%	95	99	94	97
Tensile strength ($\times 10^3$)/psi	55	75	65	160
Yield strength ($\times 10^3$)/psi	39	50	25	—
Elongation/%	25	25	35	4
Hardness, RB	55	85	45	RC 20

4. Design Parameters

For small, complex parts, MIM offers the designer a freedom of geometry similar to that for thermoplastic injection moulding or investment casting. As with any process, however, there are limits of applicability. The following paragraphs will establish the practical range for a number of design characteristics.

(1) Size Range

MIM processing is best suited for relatively small parts. On a volume basis, parts should typically be smaller than a tennis ball, with the process being most cost effective when parts are smaller than a golf ball. Irregularly shaped parts can be produced with a major axis up to about 4 inches (102mm) in length. The lower limit of the size range is determined only by the limitations of the injection moulding process itself. Parts as small as 0.25×0.10×0.05 inches (6.4mm×2.5mm×1.3mm) are economically feasible. The upper limit of part size is established by processing economics, not technical limitations. Larger parts can be made, but may not be cost effective. As part sizes increase, the powder cost rapidly becomes a significant percentage of the overall part cost.

(2) Weight Range

Obviously, the weight of a part isrelated to its size and section thicknesses. Part weights typically range from 0. 1 grams to 150 grams. However, the process is most economical for parts in the range of 1 to 25 grams.

(3) Density/Porosity

The density of sintered MIM parts is typically greater than 94% of full density. Depending on the material and processing schedule used, densities approaching theoretical may be obtained. However, MIM is generally not offered as a "full density" process. Density enhancement can be achieved through subsequent hot isostatic pressing (HIP) should it be required.

(4) Tolerances

MIM processing normally requires a dimensional tolerance of ±0. 3%. As part size decreases, increasingly tighter total tolerances can be achieved as would be expected. However, the reduction in tolerances is not directly proportional to decreasing dimensions and may depend on material, part geometry and process requirements. A tolerance of ±0. 1% can generally be held on a small, selected dimension when the mould has been "fine-tuned".

(5) Surface Finish

Surface finish on MIM parts is approximately 32 rms, significantly better than most investment castings. While the surface finish might be expected to be enhanced through the use of very fine starting powders, studies have shown that the degree of surface smoothness is heavily dependent on the sintering operation and is relatively independent of the powder particle size and finish of the mould cavities. MIM has replaced both investment cast and die cast parts in situations where extensive polishing was required on the castings in order to achieve an improved surface finish.

(Source: Merhar J. *Overview of metal injection moulding*. Metal Powder Report, 1990, 45(5))

Words and Expressions

embryonic　*a.* 胚胎的，似胚胎的
feedstock　*n.* 原料，给料
polyethylene　*n.* 聚乙烯
polypropylene　*n.* 聚丙烯
ingredient　*n.* 成分，要素
binder　*n.* 黏合剂
toothpaste　*n.* 牙膏
cavity　*n.* 洞，腔，（牙齿的）龋洞
extraction　*n.* 取出，提炼，萃取
sinter　*v.* ［冶］烧结，热压结
batch　*n.* 一批，一批量
shrinkage　*n.* 收缩，减低
supplementary　*a.* 补充的，附加的，（角）互补的
assembly　*n.* 装配，程序集，集会
reproducibility　*n.* 再现性，再生性
orthodontic　*a.* 畸齿矫正的，齿列矫正的
proportional　*a.* 成比例的，相称的，协调的
smoothness　*n.* 平滑，柔滑，平坦
injection moulding　注射成形
methyl cellulose　甲基纤维素
draft angle　［机］模锻斜度，拔模角度
green strength　［材］湿压强度，生坯强度
silicon steel　硅钢
surface finish　表面抛光，表面光洁度

16.3 问题与讨论

16.3 Questions and Discussions

(1) What is casting process?

(2) The difference between shape casting and ingot casting.

(3) What is the driving force of solidification?

(4) Why the nuclei has to grow beyond a critical size?

(5) What is homogeneous/heterogeneous nucleation?

(6) Three regions during solidification form surface to core.

单元 17　回复、再结晶与晶粒长大
Unit 17　Recovery, Recrystallization and Grain Growth

扫码查看
讲课视频

17.1　教学内容
17.1　Teaching Materials

Recrystallization has been identified as a process in metallic solids since the last part of the nineteenth century, when it was supposed that cold working of a metallic workpiece destroyed its crystallinity and that subsequent heating restored the crystalline nature by a process then naturally coined with the name "recrystallization". Nowadays we would define recrystallization as a process that leads to a change of the crystal orientation (distribution) for the whole polycrystalline specimen, in association with a release of the stored strain energy as could have been induced by preceding cold work. Recrystallization restores the properties as they were before the cold deformation. Recrystallization (and recovery and grain growth) occurs in all types of crystalline materials, so not only in metals. The industrial need for understanding the effects of deformation in material forming production steps and of subsequent annealing processes is obvious. Then it may come as a surprise that even about 150 years of research in this area have not led to comprehensive models describing these processes on the basis of fundamental insight such that reliable application for technological purposes can be guaranteed. One of the main reasons for this deficiency is undoubtedly our still limited understanding of the plastically deformed state.

1. Recovery

The defects introduced by plastic deformation processes, as cold rolling, and of importance in subsequent recovery and recrystallization processes, are predominantly dislocations. Point defects, as vacancies, are also introduced upon plastic deformation. In particular if the stacking fault energy is relatively low, dissociation of the dislocations occurs, cross-slip is hindered and twinning becomes a preferred mode of plastic deformation. Also, if not enough slip systems are available, as can occur with hexagonal metals, the initial plastic deformation can occur by slip (dislocation glide), but deformation twinning can become important upon progressing plastic deformation.

Recovery, as induced by annealing after plastic deformation, leads to a change of the dislocation microstructure and thereby a partial restoration of the material properties as before the plastic deformation is realized. It should be remarked that recovery processes can also operate in materials containing dislocations and non-equilibrium amounts of point defects (as vacancies) and which have not been subjected to pronounced plastic deformation by the exertion of external mechanical loads. In this last case recovery can restore fully the original material properties.

During the rearrangement/partial annihilation of the dislocations in the process of recovery, the grain boundaries in the material do not move; the recovery process occurs more or less homogeneously throughout the material, in flagrant contrast with recrystallization, characterized by the sweeping of high-angle grain boundaries through the deformed matrix, which process thus takes place explicitly heterogeneously.

2. Recrystallization

The heterogeneous formation of new, strain-free grains growing, by a migrating high-angle grain boundary, into the deformed matrix typifies the recrystallization process (Figure 1). This immediately indicates the driving force for recrystallization: the complete release of the strain energy induced by the preceding process of cold work and as remaining after the subsequent recrystallization-foregoing recovery.

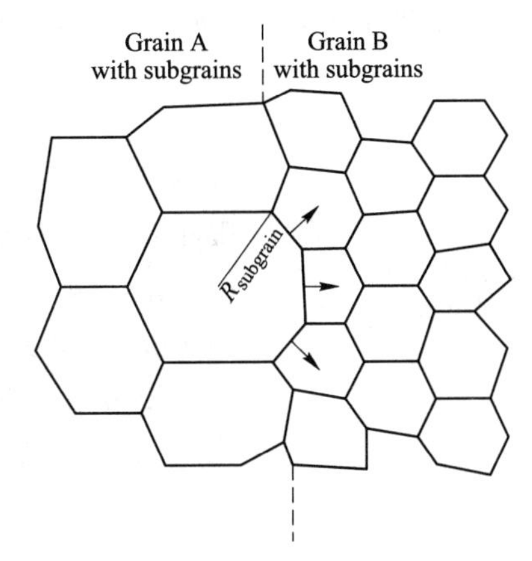

Figure 1 Grains A and B exhibit a polygonized dislocation cell (subgrain) structure
(If a subgrain, located at the A/B (high-angle) grain boundary, is large enough it can act as a location for the initiation of recrystallization)

What then are viable mechanisms for initiating recrystallization? If genuine nucleation of a strain-free grain, separated by a mobile high-angle grain boundary from the deformed matrix, is impossible, it appears natural to look for regions in the deformed microstructure, the growth of which would lead to a reduction of the stored energy in the specimen. In other words, the heterogeneity of the deformed microstructure may provide the key to the initiation of the recrystallization process. Strain-induced grain-boundary migration is thought to be initiated at a high-angle grain boundary in the deformed microstructure where the dislocation density at both sides of the boundary is significantly different due to the previous (cold) work, which can be a consequence of the dependence on crystal orientation of a grain to applied external loading. Finally it is remarked that subgrain coarsening, in the bulk of a polygonized/dislocation cell structured grain, can be a precursor for the initiation of recrystallization. Such subgrain coarsening is dominated by the migration of low-angle boundaries (the boundaries of the subgrains), which by itself is no recrystallization.

3. Grain Growth

After completion of the recrystallization process as discussed in the above section, a coarsening of the microstructure can occur, driven by the release of grain-boundary energy: the larger grains grow at the expense of the smaller grains. As the driving force for this process is distinctly smaller than for recrystallization, the velocity of the migrating grain boundaries is smaller than in the case of recrystallization. Two cases of grain growth can be discerned.

(1) Normal grain growth, characterized by an approximately uniform velocity for the migrating grain boundaries throughout the specimen, with the consequence that the grain size remains more or less uniform throughout the specimen;

(2) Abnormal grain growth, characterized by mobile grain boundaries for only a few grains, with the result that these few grains become very large as compared to the remaining majority of the grains. This last process has, confusingly, also been called secondary recrystallization.

(Source: Mittemeijer. *Fundamentals of Materials Science*. Berlin: Springer, 2010)

Words and Expressions

orientation　*n.* 取向
recovery　*n.* 回复
deficiency　*n.* 缺乏，不足，缺陷
cross-slip　*n.* 交滑移
hexagonal　*a.* 六角形的
annihilation　*n.* 毁灭，湮灭
initiation　*n.* 开始，萌生，引发
flagrant　*a.* 鲜明的，不能容忍的，非常的
twinning　*n.* 孪生，成对，双晶形成
cold deformation　冷加工，冷变形
grain growth　晶粒长大
slip system　滑移系

Phrases

identified as　定义为……，确定为……
coined with　产生，创造
define as　定义为……
in association with　与……相关
release of　释放
look for　寻找，指望，寻求
on the basis of　根据，基于……
in other words　换句话说
driven by　由……驱动

Notes

(1) Recrystallization has been identified as a process in metallic solids since the last part of the nineteenth century, when it was supposed that cold working of a metallic workpiece destroyed its crystallinity and that subsequent heating restored the crystalline nature by a process then naturally coined with the name "recrystallization".

it was supposed 译为“本想……，应该认为……，预期……”；cold working 译为“冷加工”；coined with 译为"产生，创造"；when 引导条件状语从句。

参考译文：自19世纪后期，再结晶被确定为固态金属的一项工艺，当人们认为冷加工破坏了金属工件的结晶度，随后的加热过程恢复了其结晶性质，然后自然产生了"再结晶"这个名称。

(2) Then it may come as a surprise that even about 150 years of research in this area have not led to comprehensive models describing these processes on the basis of fundamental insight such that reliable application for technological purposes can be guaranteed.

lead to 译为"导致，致使"；on the basis of 译为"根据；基于……；such that 译为"如此……以至于"，其中 that 引导 such 同位语从句。

参考译文：然而令人惊讶的是，即使在这一领域进行了大约150年的研究，也没有形成基于基本理解建立描述这些过程的综合模型，从而保证技术上的可靠应用。

(3) The defects introduced by plastic deformation processes, as cold rolling, and of importance in subsequent recovery and recrystallization processes, are predominantly dislocations.

recovery(回复)和 recrystallization(再结晶)是金属在回火过程中发生的内部结构变化过程；cold rolling 译为"冷轧"，是金属加工的一种重要手段。

参考译文：冷轧等塑性变形过程中引入的缺陷主要是位错，在随后的回复和再结晶过程至关重要。

(4) Recovery, as induced by annealing after plastic deformation, leads to a change of the dislocation microstructure and thereby a partial restoration of the material properties as before the plastic deformation is realized.

annealing 译为"退火"，也是热处理常用工艺之一；plastic deformation 译为"塑性变形"。

参考译文：塑性变形后退火引起的回复过程，会导致位错微观结构变化，从而使材料性能部分恢复到塑性变形之前的状态。

(5) The recovery process occurs more or less homogeneously throughout the material, in flagrant contrast with recrystallization, characterized by the sweeping of high-angle grain boundaries through the deformed matrix, which process thus takes place explicitly heterogeneously.

homogeneously 与 heterogeneously 构成反义词；high-angle grain boundaries 译为"大角度晶界"；take place 译为"发生，产生"。

参考译文：与再结晶形成鲜明对比的是，整个材料的回复过程或多或少都是均匀的。再结晶的特征是通过变形基体扫过大角度的晶界，这一过程明显是不均匀的。

(6) If genuine nucleation of a strain-free grain, separated by a mobile high-angle grain boundary from the deformed matrix, is impossible, it appears natural to look for regions in the deformed microstructure the growth of which would lead to a reduction of the stored energy in the specimen.

if 引导一个条件状语从句；strain-free grain 指"无畸变的晶粒"。

参考译文：如果一个无畸变的晶粒，被一个移动的大角度晶界从变形基体中分离出来，不可能真正成核，那么在变形组织中寻找能导致试样中储存能量减少的区域似乎是很自然的。

17.2　阅读材料

17.2　Reading Materials

Diffusion in Metals and Alloys

1. Introduction

Transport in materials via random individual atomic migration steps (jumps) is called "diffusion". General understanding and detailed theory of diffusional transport represent a classical field of material science. Measurement of diffusion in solids needs sensitive techniques and well-developed microscopy. Not surprising that a scientific proof of diffusion in solid metals came rather late in history of science. Although diffusion in solid state is slow, knowledge of diffusion processes is nevertheless fundamental to material scientists and engineers. Tailoring of microstructures by solid-state reactions is largely based on diffusion processes at elevated temperatures. Long-term stability and aging of materials, especially important in high-temperature applications, are controlled by diffusion. With downscaling of structural dimensions in current technology, control of diffusional transport on the nanometer scale gets decisive in electronic devices or MEMS (Micro Electro-Mechanical Systems), even in nanocrystalline construction materials for operation close to room temperature.

Understanding diffusion from a theoretical point of view is challenging as many different fields are involved. Complex boundary value problems of continuum diffusion equations found beautiful mathematic solutions in the past. A statistical description of independent atomic jumps requires profound thermodynamics making use of equilibrium and nonequilibrium concepts. It was none other than Albert Einstein who provided the missing link between the continuum description of diffusion and the random walk of individual atoms. Beginning with the second half of past century, the microscopic mechanisms of atomic jumps attracted increasing attention. Nowadays this field is particularly driven by the possibilities of computer simulation. Energetics and kinetics of atomic jumps are calculated by ab initio method, while the whole diffusion process is simulated by Monte Carlo (MC) or to greater detail by Molecular Dynamic (MD) simulation. From the experimental point of view, one may state that the bulk of required diffusion data for metals are already known. But confronted to the actual task of discussing a given reaction or material, many scientists are suddenly surprised that just in their particular case required diffusion data are still missing. Continuous development of alloys and intermetallic phases motivates further new and more accurate measurement of respective diffusion properties. This is particularly true in the case of short circuit transport on the nanoscale. Coefficients of grain boundary diffusion in the dilute limit are partly known, but interdiffusion within grain boundaries in the technically interesting high concentration range, properties of interphase boundaries, triple junctions or higher order defects, and transport barriers at interfaces are still widely unexplored landscape.

2. Atomistic Mechanism and Fundamental Relations of Diffusion

(1) Elementary Diffusion Mechanisms in Metals

Crystal structures of metals are distinguished by their dense packing. Hard to imagine, how atoms may exchange their lattice sites in a dense lattice. It seems to be natural and has been confirmed by experiment that diffusional transport in solid metals can only appear by point defects that act as vehicles of transport. The few possibilities of defect-mediated atomic migration that are realized in metals are illustrated in Figure 1. In real experiment, one can mark a tiny fraction of atoms by another isotope, possibly a radioactive one. Atoms marked in this way are called tracer atoms.

The migration of small solutes dispersed on interstitial sites of the host represents conceptually the simplest case of diffusion [Figure 1(a)]. Provided low concentration, a solute on an interstitial site finds its neighbor interstices usually empty and so a spontaneous jump to the next free site is always possible. Since elastic energy required for distortion increases with the square of the solute size, only rather small atoms, such as hydrogen, carbon, or oxygen, are candidates to occupy interstitial sites in thermal equilibrium and to perform easy diffusion jumps.

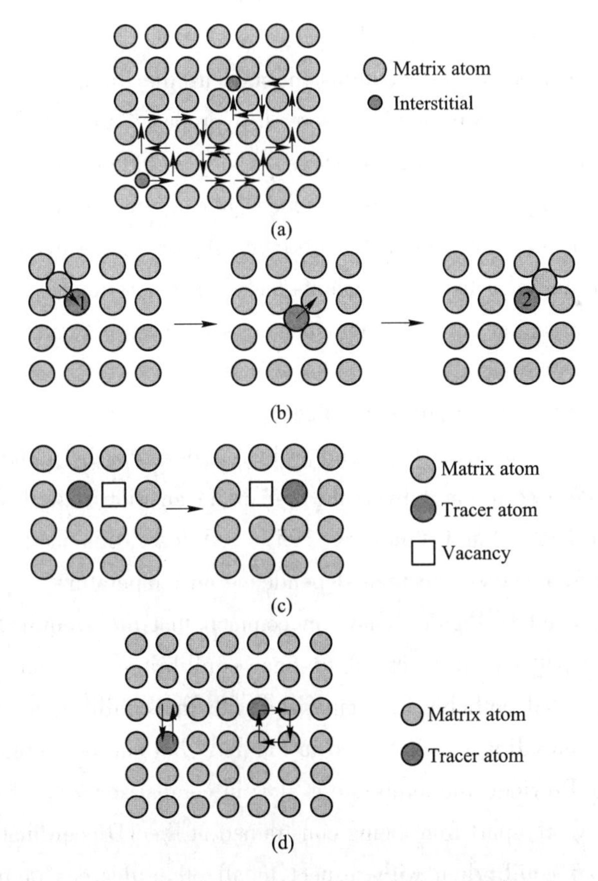

Figure 1 Site exchange in crystalline metals

(a) Interstitial diffusion; (b) Interstitialcy mechanism; (c) Vacancy mechanism; (d) Hypothetical direct site exchange (left) and ring exchange (right)

By contrast, misplacing a host atom itself into an interstitial site (self-interstitial) requires in metals so large amount of energy, that a measurable fraction of self-interstitials is only expected after irradiation or severe plastic deformation. If self-interstitials are nevertheless present (e. g. as a consequence of irradiation with energetic particles), they allow an indirect transport mechanism, usually termed as the interstitialcy mechanism [Figure 1(b)]. In the first step, a substitutional atom is kicked by a neighbored interstitial from its lattice position (1) into the next free interstice, while the former interstitial takes the original lattice site. The newly formed interstitial repeats this step by substituting a third lattice atom, and so finds itself at the end at the new lattice position (2). In contrast to self-interstitials, vacant lattice sites do belong to thermal equilibrium of a crystal in considerable concentrations. Near to the melting point, one may even expect a fraction of 10^4 to 10^3 of unoccupied lattice sites. Therefore, site exchange of atoms with neighbored vacancies [Figure 1(c)] represents the most common mechanism of self or (substitutional) solute diffusion in metals. In comparison with small interstitials, this vacancy-mediated diffusion is rather slow, since atoms have to wait for a vacancy appearing in its neighborhood, before a jump becomes possible. Intrinsic vacancies are formed by thermal activation. Therefore, temperature dependence of this kind of diffusion is quite pronounced.

Alternative cooperative mechanisms of diffusion, such as the direct pair exchange or a closed ring replacement sequence [Figure 1(d)] were also proposed initially to explain diffusion in crystalline metals. However, direct pair exchange in dense lattices requires too much elastic distortion and the probability for correlated sequences like a ring exchange turned out to be too small for being significant in crystalline metals. In contrast, convincing evidence has been gathered by computer simulation and also by measurements based on the isotope effect to the diffusion rate that the ring exchange or similar correlated replacement sequences play an important role in liquids and metallic glasses.

(2) The Individual Jump: Transition State Theory

Understanding diffusion rates needs at first hand an idea to describe the probability of an individual jump. Quite a reasonable description delivers the so-called transition state theory which is based on classical work by Wert (1950) and Vineyard (1957). At least, this model provides a practical understanding of jump frequencies and their dependence on temperature.

The mechanisms sketched in Figure 1 have in common that they require the jumping atom to pass a saddle on their pathway from initial to final equilibrium position. Obtaining the saddle position requires an elevated enthalpy in comparison with the equilibrium positions. The pathway itself may be a complex curvilinear as indicated in Figure 2, but can be certainly projected to a one dimensional coordinate. Provided the jump rate is slow in comparison with phonon frequencies, one may assume that the crystal, apart from being constrained at the 1D coordinate along the transition path, establishes thermal equilibrium with respect to all other degrees of freedom of the phase space. Under this assumption, the jump rate w between two selected lattice sites can be estimated by the thermodynamic probability of finding the jumping atom at the saddle times its Maxwell equilibrium velocity along the transition path (Schoeck, 1980). As most compact descriptions, the theory arrives at (Vineyard, 1957)

$$\omega = \Gamma_0 \exp\left(-\frac{H^M}{k_B T}\right) \tag{1}$$

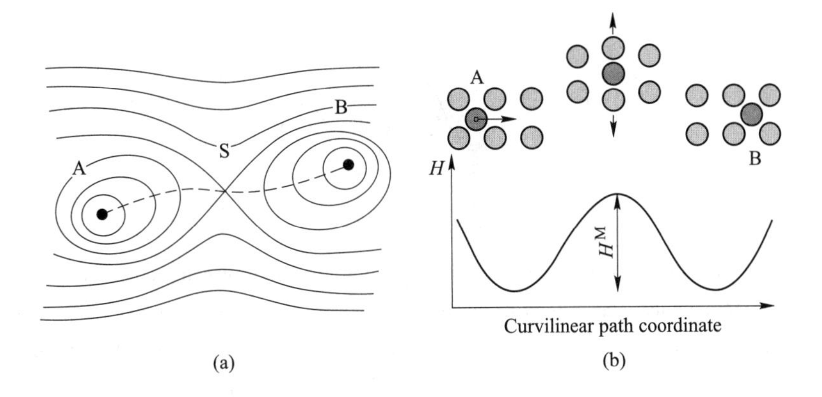

Figure 2　Energetic situation of the atomic jump

(a) 2D representation of the enthalpy landscape between two equilibrium positions (A and B) of the jumping atom. Both positions are linked by a curvilinear pathway;

(b) Along this path, a clearly defined energy maximum appears at the saddle position. The respective migration enthalpy HM may be imagined as elastic energy needed to push outward neighbored atoms that hinder the transition

(3) Continuum Diffusion Equations

A one-dimensional scheme as sketched in Figure 3 is sufficient to obtain the desired fundamental relations. Imagine two adjacent planes of interstices spaced by the jump distance l. Let them be occupied by different numbers $n(x)$ and $n(x+\lambda)$ of interstitial atoms per area being able to jump. The enthalpy landscape of an atom constrained to the diffusional pathway is sketched in the bottom row of the figure.

A few solutions of the second Fick's law in its linear form $\frac{\partial c}{\partial t} = D\frac{\partial^2 c}{\partial x^2}$ are so important that they must be discussed here. On the one hand, they form the basis for the experimental determination of diffusion coefficients and on the other hand they are generic to derive further more specialized solutions.

If composition profiles are investigated on the length scale of micrometers in a macroscopic sample of millimeter dimensions. For this case, major solutions are based on thebell-shaped Gaussian function which broadens in time but always vanishes far from the diffusion zone:

$$c(x,t) = \frac{m}{2\sqrt{\pi D t}} \exp\left(-\frac{x^2}{4Dt}\right) \tag{2}$$

In the presented form, Equation (2) is normalized so that, when integrating along the x-axis, the total amount of diffusing species amounts to m. Furthermore, at the very beginning ($t \to 0$), this function is

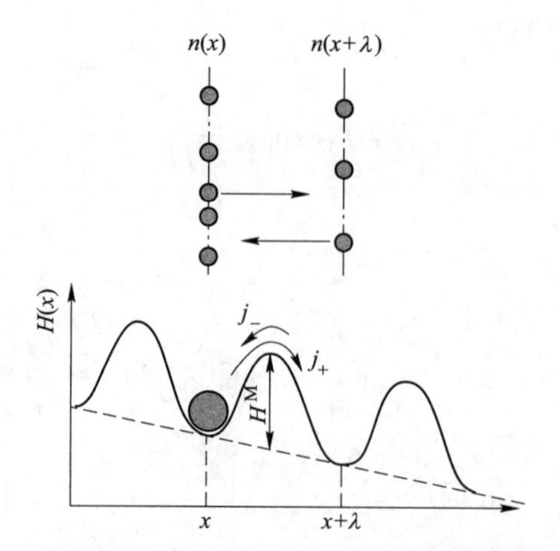

Figure 3 One-dimensional enthalpy profile assuming a constant driving force. Two adjacent layers of interstitial sites spaced by a distance l are distinguished by local energy minima between which a barrier has to be overcome. Owing to a driving force, the barrier from left to the right gets slightly lower than that from right to left

a representation of Dirac's δ function, meaning that all diffusing species are concentrated first in an infinitely thin slab and then diffuse outward to both sides [Figure 4(a)].

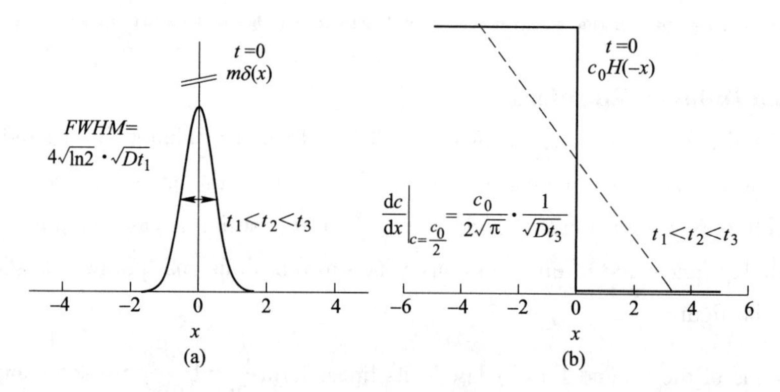

Figure 4 Evolution of the concentration profile starting

(a) From an infinitesimally thin layer of the diffusor; (b) From a stepwise profile (A/B diffusion couple)

The most important example is the thick film solution to be used for a diffusion couple that comprises two half spaces initially of different composition [Figure 4(b)]. The solutions is usually expressed as:

$$c(x) = \frac{c_0}{2}\left[1 - \mathrm{erp}\left(\frac{x}{2\sqrt{Dt}}\right)\right] \tag{3}$$

which inherits from the Gaussian distribution the characteristic $\sqrt{Dt}$ dependence of the width of the mixed zone around the initial welding plane. Besides interdiffusion couples, the same solution is also useful in the description of diffusion into a "half space" sample (located on the positive x-axis) keeping the surface concentration (at $x=0$) at the constant level $c_0/2$. The solution further plays a role in the quantitative description of short circuit transport along grain boundaries.

(Source: David E laughlin, kazuhiro Hono. *Physical Metallurgy*. Amsterdam: Elsevier, 2015)

Words and Expressions

downscaling *v.* 缩小……的尺寸，降低……的程度
coefficient *n.* [数] 系数
interdiffusion *n.* [物] 相互扩散
isotope *n.* 同位素
radioactive *a.* 放射性的，有辐射的
irradiation *n.* 照射，发光，放射
interstice *n.* 间隙，缝
interstitialcy *n.* 推填子，间隙原子
probability *n.* 可能性；[数] 概率
curvilinear *a.* 曲线的，由曲线组成的
enthalpy *n.* [热] 焓，热含量
ab initio （拉丁）从头开始，从头算
dilute limit 稀释极限
tracer atom 示踪原子

17.3 问题与讨论

17.3 Questions and Discussions

(1) What is recrystallization process and what is the driving force?
(2) What will happen during recovery process?
(3) If not enough slip systems are available, what will happen during plastic deformation?
(4) What then are viable mechanisms for initiating recrystallization?
(5) The driving force and two cases of grain growth.

第五部分　金属材料的性能

Part 5　Properties of Metallic Materials

单元 18　金属与合金的机械性能
Unit 18　Mechanical Properties of Metals and Alloys

扫码查看
讲课视频

18.1　教学内容
18.1　Teaching Materials

The deformation of a solid may be constituted of change in size or volume, called dilation, or change in shape, known as distortion. Hence, mechanical deformation can be defined as a change in the size or shape of an object due to an applied force, which may be tensile or pulling, compressive, shear, bending or torsion (twisting). When deformation occurs, the atoms are displaced from their equilibrium interatomic spacing. This atomic displacement causes to develop internal interatomic attractive force or repulsive force that opposes the applied force. If the applied force is not too large, the interatomic force may be sufficient to completely resist the applied force, allowing the object to return immediately to its original state on complete release of the applied load. This type of instantaneous self-reversing deformation is called recoverable or elastic deformation. A larger applied force may lead to a permanent set or plastic deformation of the material or even to its structural failure. In general, if a material exhibits the ability to undergo plastic deformation under load it is ductile, otherwise it is brittle.

1. Tension

The static tensile test performed under uniaxially applied load is the most extensively used experimental test method which characterizes many important mechanical properties of materials. From this test, one can know the material's elastic properties such as elastic limit, proportional limit, modulus of elasticity and modulus of resilience in tension. For engineering purposes, most commonly used inelastic properties such as yield strength, tensile strength, breaking or fracture strength, ductility and toughness can also be measured from this test record.

2. Compression

Compression test is simply the reverse of the tension test with reference to the sense or direction of applied stress. There are many similarities between static compression and tension test as far as the behaviour of materials is concerned. However, static compression differs adequately from static tension in some important matters, such as (1) manner of loading, (2) test specimen geometry, (3) frequent differences in stress-strain diagrams with a pronounced difference in the plastic range, (4) the extent of ductility shown by a material, and (5) the mode of failure and fracture surface appearance. Compression test is preferable to tension test for brittle materials.

3. Hardness

The term "hardness" is a structure-sensitive mechanical property of materials, primarily associated with the surface. In general, the hardness is defined as the resistance of a material to permanent or plastic deformation of its surface, usually by indentation, under static or dynamic load. The hardness test becomes the easiest way of measuring strength property. Moreover, the quality level of materials or products may be checked or controlled by the hardness measurements. Often results of such measurements become a determining factor in the acceptance or rejection of products, especially in any heat-treating or hardening process.

4. Bending

Majority of structural members used in engineering applications are subjected to some bending. The reaction of the member in bending is to develop internal stresses that will oppose its extension and compression. A beam is a suitably supported structural member that has length reasonably higher than its lateral dimensions and undergoes bending under the application of transverse loads. A beam under transverse loading will be subjected to bending moment which will induce stresses in the beam, and these stresses are called bending stresses.

5. Torsion

The twisting action of one section of a member with respect to an adjoining section is called torsion. Since torsion involves pure shear and shear stresses cause plastic flow, the torsion test is performed for theoretical studies of plastic deformation process. Torsion tests are carried out to determine properties of materials, such as (1) the modulus of elasticity in shear, (2) the torsional yield strength, and (3) the ultimate torsional shear strength known as the modulus of rupture. But the torsion test cannot be applied to determine the shearing strength of brittle materials like cast iron.

6. Impact Loading

Many structures and machines or their parts are subjected to dynamic loads. There are two important types of dynamic loading. One of them includes the application of rapidly fluctuating loads as occurred in fatigue phenomenon. The other one involves the sudden application of load, as from the impact of a moving mass. The stress produced by the impact loading depends upon the amount of energy that is expended to cause deformation of the parts receiving the blow.

7. Creep and Stress Rupture

If, under any conditions, deformation extends over a period of time when the load or stress is kept constant, this time-dependent permanent deformation under constant load or stress is called creep and is observed in both crystalline and amorphous materials. Initially creep deformation may be small, but over the lifetime of the structure deformation can grow large and even result in final fracture without any increase in load. The higher the temperature, the more pronounced is the creep phenomenon.

8. Fatigue

Most structural assemblies are subjected to fluctuating or repetitive stresses of sufficient magnitude for sufficient number of times, although the maximum value of this applied fluctuating stress may be considerably less than the static tensile strength of the material. Such condition of dynamic loading produces a permanent damage to the material that leads to failure after a considerable period of service. This progressive failure of the material at a stress much lower than that required to cause fracture on a single application of load is called a fatigue failure. It has been estimated that 80%~90% of all mechanical failures in service are due to fatigue.

(Source: Amit Bhadun. *Mechanical Properties and Working of Metals and Alloys*. Berlin: Springer, 2018)

Words and Expressions

dilation　*n.* 膨胀，扩大
distortion　*n.* 变形
uniaxially　*ad.* 单向地，单轴地
lateral　*a.* 侧面的，横（向）的
indentation　*n.* 压痕，刻痕，凹陷
torsion　*n.* 扭转
twisting　*n.* 快速扭转，缠绕
interatomic spacing　原子间距
attractive force　吸引力
repulsive force　斥力
elastic deformation　弹性变形
yield strength　屈服强度
bending stress　弯曲应力
brittle fracture　脆性断裂
stress rupture　应力断裂
fatigue failure　疲劳断裂，疲劳失效

Phrases

be constituted of　由……组成
be sufficient to　足矣……，对……来说足够了
with reference to　关于，有关，根据
as far as　至于，直到，就……而言
differ from　与……不同；区别于……

Notes

(1) Mechanical deformation can be defined as a change in the size or shape of an object due to an applied force, which may be tensile or pulling, compressive, shear, bending or torsion (twisting).

defined as 译为“定义为”；applied 译为“施加的，应用于”；“which”引导定义从句，修饰 applied force。

参考译文：机械变形可以定义为由一个施加的外力引起的物体的大小或形状的变化，可能是拉伸、压缩、剪切、弯曲或扭转（扭曲变形）。

(2) The static tensile test performed under uniaxially applied load is the most extensively used experimental test method which characterizes many important mechanical properties of materials.

which 引导定语从句，修饰“test method”。

参考译文：在单轴下施加载荷进行的静态拉伸测试是使用最广泛的测试方法，它表征了材料的许多重要的机械性能。

(3) Compression test is simply the reverse of the tension test with reference to the sense or direction of applied stress. There are many similarities between static compression and tension test as far as the behaviour of materials is concerned.

compression(压缩) 与 tension(拉伸) 构成反义词；with reference to 译为“关于，依照”。

参考译文：根据所施加应力的方向，压缩测试只是拉伸测试的反向。就材料的性能而言，静态压缩试验和拉伸试验之间有许多相似之处。

(4) In general, the hardness is defined as the resistance of a material to permanent or plastic deformation of its surface, usually by indentation, under static or dynamic load.

permanent deformation 译为“永久变形”；plastic deformation 译为“塑性变形”。static 与 dynamic 互为反义词。

参考译文：一般来说，硬度定义为材料在静态或动态载荷下抵抗表面永久或塑性变形（通常是压痕）的能力。

(5) Torsion tests are carried out to determine properties of materials, such as (1) the modulus of elasticity in shear, (2) the torsional yield strength, and (3) the ultimate torsional shear strength known as the modulus of rupture.

torsion 与 tension 和 compression 是材料受力的主要方式；yield strength 译为“屈服强度”；shear strength 译为“剪切强度”。

参考译文：扭转试验是为了确定材料的特性而进行的，例如（1）剪切弹性模量，（2）扭转屈服强度和（3）称为断裂模量的极限扭转剪切强度。

(6) One of them includes the application of rapidly fluctuating loads as occurred in fatigue phenomenon. The other one involves the sudden application of load, as from the impact of a moving mass.

one…, the other…两个并列句子，一个表述了 fatigue(疲劳)，另一个表述了 impact(冲击)。

参考译文：其中之一包括施加在疲劳过程中快速波动的载荷；另一个涉及由于移动质量的冲击而突然施加的载荷。

18.2　阅读材料

18.2　Reading Materials

Strengthening Mechanisms of Aluminum

There are five separate strengthening mechanisms that can be applied to the aluminium alloys. These are grain size control, solid solution alloying, second phase formation, strain hardening (cold work) and precipitation or age hardening.

1. Grain Size Control

Grain size is not generally used to control strength in thealuminium alloys, although it is used extensively in reducing the risk of hot cracking and in controlling both strength and notch toughness in C/Mn and low-alloy steels. In general terms, as grain size increases, the yield and ultimate tensile strengths of a metal are reduced. The yield strength σ_y, is related to the grain size by the Hall-Petch equation: $\sigma_y = \sigma_1 + k_y d^{-1/2}$, where d is the average grain diameter, and σ_y and σ_I are constants for the metal. Typical results of this relationship are illustrated in Figure 1.

In the aluminium alloys the strength loss due to grain growth is a marginal effect, with other effects predominating. Grain size does, however, have a marked effect on the risk of hot cracking, a

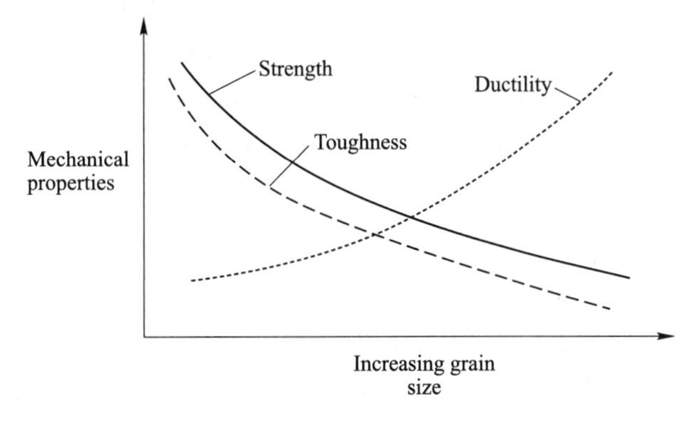

Figure 1　General relationship of grain size with strength, ductility and toughness

small grain size being more resistant than a large grain size. Titanium, zirconium and scandium may be used to promote a fine grain size, these elements forming finely dispersed solid particles in the weld metal. These particles act as nuclei on which the grains form as solidification proceeds.

2. Solid Solution Strengthening

Very few metals are used in the pure state, as generally the strength is insufficient for engineering purposes. To increase strength the metal isalloyed, that is mixed with other elements, the type and amount of the alloying element being carefully selected and controlled to give the desired properties. Depending upon the metals involved a limit of solid solubility may be reached. Microscopically a solid solution is featureless but once the limit of solid solubility is reached a second component or phase becomes visible. This phase may be a secondary solid solution, an intermetallic compound or the pure alloying element. The introduction of a second phase results in an increase in strength and hardness, for instance iron carbide (Fe_3C) in steels, copper aluminide ($CuAl_2$) in the aluminium-copper alloys and silicon (Si) in the aluminium-silicon alloys.

In solid solution alloying the alloying element or solute is completely dissolved in the bulk metal, the solvent. There are two forms of solid solution alloying—interstitial and substitutional—illustrated in Figure 2. Interstitial alloying elements fit into the spaces, the interstices, between the solvent atoms, and substitutional elements replace or substitute for the solvent atoms, provided that

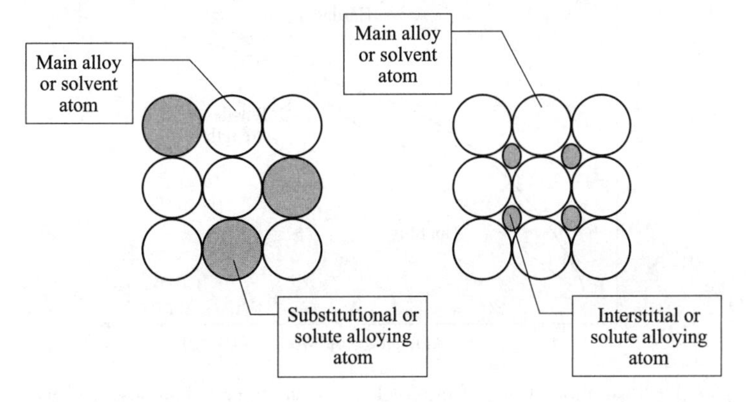

Figure 2　Schematic illustration of substitutional and interstitial alloying

the diameter of the substitutional atom is within ±15% of the solvent atomic diameter. The effect of these alloying elements is to distort the space lattice and in so doing to introduce a strain into the lattice. This strain increases the tensile strength but as a general rule decreases the ductility of the alloy by impeding the slip between adjacent planes of atoms. Many elements will alloy with aluminum but only a relatively small number of these give an improvement in strength or weldability. The most important elements are silicon, which increases strength and fluidity; copper, which can give very high strength; magnesium which improves both strength and corrosion resistance; manganese, which gives both strength and ductility improvements; and zinc, which, in combination with magnesium and/or copper, will give improvements in strength and will assist in regaining some of the strength lost when welding.

3. Cold Working or Strain Hardening

Cold work, work hardening or strain hardening is an important process used to increase the strength and/or hardness of metals and alloys that cannot be strengthened by heat treatment. It involves a change of shape brought about by the input of mechanical energy. As deformation proceeds the metal becomes stronger but harder and less ductile, as shown in Figure 3, requiring more and more power to continue deforming the metal. Finally, a stage is reached where further deformation is not possible—the metal has become so brittle that any additional deformation leads to fracture. In cold working one or two of the dimensions of the item being cold worked are reduced with a corresponding increase in the other dimension(s). This produces an elongation of the grains of the metal in the direction of working to give a preferred grain orientation and a high level of internal stress. The increase in internal stress not only increases strength and reduces ductility but also results in a very small decrease in density, a decrease in electrical conductivity, an increase in the coefficient of thermal expansion and a decrease in corrosion resistance, particularly stress corrosion resistance. The amount of distortion from welding is also likely to be far greater than from a metal which has not been cold worked.

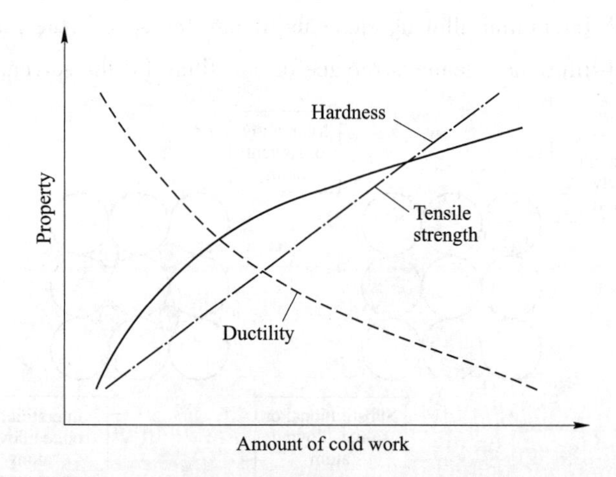

Figure 3　Illustration of the effect of cold work on strength, hardness and ductility

If a cold worked metal is heated a temperature is reached where the internal stresses begin to relax and recovery begins to take place. This restores most of the physical properties of the unworked metal but without any observable change in the grain structure of the metal or any major change in mechanical properties. As the temperature is increased, recrystallisation begins to occur where the cold worked and deformed crystals are replaced by a new set of strain-free crystals, resulting in a reduction in strength and an increase in ductility. This process will also result in a fine grain size, perhaps finer than the grain size of the metal before cold working took place. On completion of recrystallisation the metal is said to be annealed with the mechanical properties of the non-cold-worked metal restored. At temperatures above the recrystallisation temperature the new grains begin to grow in size by absorbing each other. This grain growth will result in the formation of a coarse grained micro-structure with the grain size depending upon the temperature and the time of exposure. A coarse grain size is normally regarded as being undesirable from the point of view of both mechanical properties and weldability.

4. Precipitation (Age) Hardening

A similar effect in metals enables the microstructure of a precipitation hardenable alloy to be precisely controlled to give the desired mechanical properties. To precipitation or age harden an alloy the metal is first of all heated to a sufficiently high temperature that the second phase goes into solution. The metal is then "rapidly" cooled, perhaps by quenching into water or cooling in still air—the required cooling rate depends upon the alloy system. Most aluminium alloys are quenched in water to give a very fast cooling rate. This cooling rate must be sufficiently fast that the second phase does not have time to precipitate. The second phase is retained in solution at room temperature as a super-saturated solid solution which is metastable, that is, the second phase will precipitate, given the correct stimulus. This stimulus is ageing, heating the alloy to a low temperature. This allows diffusion of atoms to occur and an extremely fine precipitate begins to form, so fine that it is not resolvable by normal metallographic techniques. This precipitate is said to be coherent; the lattice is still continuous but distorted and this confers on the alloy extremely high tensile strength. If heating is continued or the ageing takes place at too high a temperature the alloy begins to overage, the precipitate coarsens, perhaps to a point where it becomes metallographically visible. Tensile strength drops but ductility increases. If the overageing process is allowed to continue then the alloy will reach a point where its mechanical properties match those of the annealed structure. Too slow a cooling rate will fail to retain the precipitate in solution. It will form on the grain boundaries as coarse particles that will have a very limited effect on mechanical properties. The structure is that of an annealed metal with identical mechanical properties. The heat treatment cycle and its effects on structure are illustrated in Figure 4.

(Source: Gene Mathers. *The welding of aluminium and its alloys*. Cambridge: Woodhead Publishing Limited, 2002)

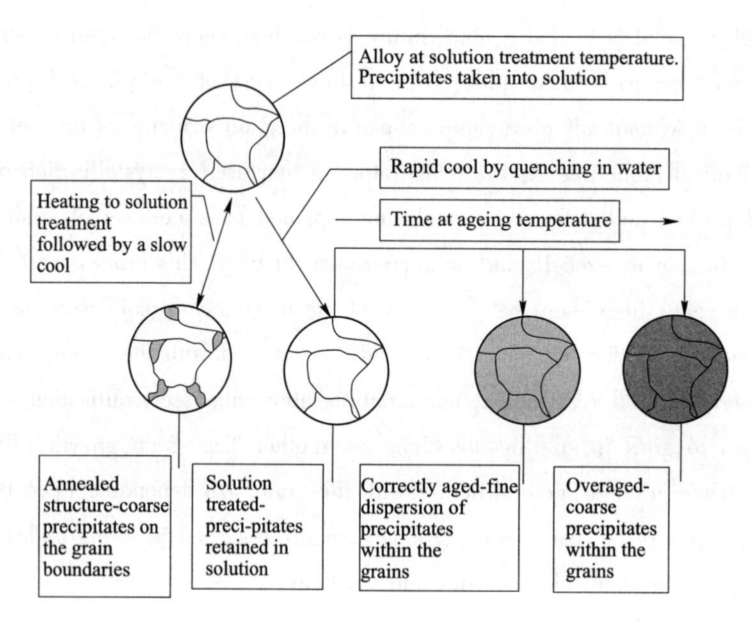

Figure 4　Illustration of the solution treatment and age-(precipitation) hardening heat treatment cycle

Words and Expressions

fluidity　*n.* 流动性，流质

scandium　*n.* ［化］钪（Sc）

metastable　*n.* 亚稳态，介稳

precipitate　*n.* 沉淀物，析出物

metallographic　*a.* 金相学的，金相的

overaging　*n.* 过时效，过老化，过度老化

hot cracking　热裂，高温龟裂

notch toughness　［力］缺口韧性

work hardening　加工硬化

marginal effect　边际效应

18.3　问题与讨论
18.3　Questions and Discussions

(1) What is mechanical deformation?

(2) The principles for elastic deformation and plastic deformation.

(3) Describe some elastic and inelastic properties.

(4) The difference between hardness and compression.

(5) Two important types of dynamic loading.

(6) What is creep or fatigue phenomenon?

(7) Methods of strengthening.

单元 19　超导性
Unit 19　Superconductivity

19.1　教学内容
19.1　Teaching Materials

1. Background

Superconductivity is a phenomenon in which the resistance of the material to the electric current flow is zero. Kamerlingh Onnes made the first discovery of the phenomenon in 1911 in mercury(Hg) as shown in Figure 1. Since that time, superconductivity has been found to occur in many metallic elements and intermetallic compounds. And more recently has been found even in organo-metallic compounds, semiconductors and ceramics. However, it is to be noted that superconductivity has not been found in all metals. For example, Li, Na, and K, have been investigated for superconductivity down to 0.08°K, 0.09°K and 0.08°K respectively and the results indicated no sign of superconductivity.

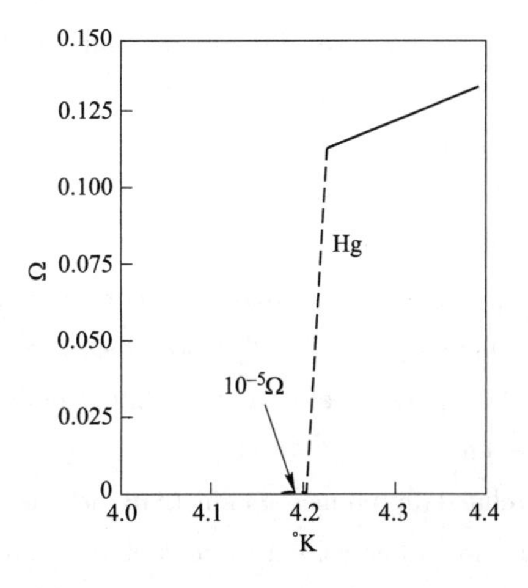

Figure 1　Showing near zero electrical resistance at about 4.2°K for mercury(Hg)
(As discovered by Kamerlingh Onnes)

Associated with the superconductive state is a perfect diamagnetism in which the magnetic flux is expelled from the material. This is known as the Meissner effect. This magnetic property associated with superconductors cannot be accounted for by the assumption that the superconducting state is equivalent to zero electrical resistance. Furthermore, experimentally, no magnetic material has been found to be a superconductor thus far.

2. The BCS Theory of Superconductivity

In 1957, Bardeen, Cooper and Schrieffer proposed a quantum theory of superconductivity, known as the BCS theory, by which they received the Nobel Prize. Their theory in non-mathematical terms can be summarized as follows.

(1) An attractive interaction between electrons through electron-lattice-electron interaction can lead to a ground state separated from excited states by an energy gap.

(2) The magnetic flux penetration depth and coherence length emerge as a consequence of the theory.

(3) It predicts a relationship between superconductivity temperature, T_c, the lattice vibration, θ_{Debye}, and average atomic mass, M, as $T_c \propto \theta_{Debye} \propto M^{1/2}$ such that $M^a. T_c = $ Constant. Since M is an average atomic mass, it is also known as the "Isotope" effect.

(4) The criterion for the transition temperature, T_c, of an element is related to the electron density of orbital $D(\varepsilon_F)$ at the Fermi surface and the electron-lattice interaction U, which can be estimated from the electrical resistivity. For $UD_{(\varepsilon_F)} \ll 1$, the BCS theory predicts $T_c = 1.14\theta \exp[-1/UD_{(\varepsilon_F)}]$.

(5) Magnetic flux through a superconducting ring is quantized and the effective unit of charge is $2e$. The BCS ground state predicts paired electrons, which are derived from quasi-free electrons.

From the time the BCS theory was proposed, a large number of experimental evidence on superconductivity has accumulated. But, they do not seem to support the theory. Some obvious paradox is: experimentally, the higher the resistivity at room temperature, the more likely that metal will be a superconductor when cooled. This directly contradicts the BCS' s idea that paired electrons originate from fermi electrons. The proposal of the Cooper electron pair is justified solely on the mathematical argument that although the kinetic energy may be high their potential energy is lowered to the extent that they become stable.

3. Experimental Facts on Superconductivity

We shall now summarize the major experimental observations made related to the superconductivity.

(1) Occurrence of superconductivity and the T_c associated with the superconducting elements have not been correlated with other physical attributes, such as atomic number, ionization potential, atomic orbital, crystal structure, etc.

(2) Superconductivity is related to normal conductivity to the extent that in general, good conductors at room temperature are either non-superconductors or poor superconductors.

(3) Magnetic materials are not superconductors; Magnetism and superconductivity appear to be mutually exclusive.

(4) Superconductivity with zero electrical resistivity does not mean it can carry an unlimited amount of current. Every superconductor has its own current limit.

(5) Superconducting current is quantized in $2e$ charges instead of one e.

(6) Superconducting current is self-sustaining in the absence of outside disturbance or interference.

(7) Based on their difference in reacting to magnetic flux, superconducting material can be

differentiated into Type-Ⅰ and Type-Ⅱ. The Type-Ⅱ superconductors are usually associated with intermetallic compounds instead of elements and their superconductivity is not easily affected even with a high magnetic field.

(8) Under high pressure, superconducting temperature of a superconductor can become higher or lower.

(9) Superconductivity has been found in metallic elements and intermetallic compounds and within their solid-solution-range. But superconductivity has not been found in an alloy with an arbitrary composition.

(10) Pure organic and inorganic material that does not exhibit any normal electrical conduction is not a superconductor.

(11) Superconductivity occurs as a low temperature phenomenon. So far, no superconductivity phenomenon has been found above 100℃K.

(12) Transition from normal conductivity to superconductivity is a virtual perfect second order phase transition; that is, there is no latent heat or a sharp finite discontinuity in the specific heat; therefore, it is a cooperative phenomenon.

These facts are extremely important in the sense that any theory proposed must agree at least qualitatively with, not one or two observations, but with all observations.

(Source: Frederick E Wang. *Bonding Theory for Metals and Alloys*. Amsterdam: Elsevier, 2005)

Words and Expressions

diamagnetism *n.* 抗磁现象
criterion *n.* 准则
quantized *a.* 量子化的；*v.* 使量子化
qualitatively *ad.* 定性地
electric current 电流
magnetic flux [电] 磁通量
quantum theory 量子理论
paired electron 成对电子
potential energy 势能
kinetic energy 动能
ionization potential 电离电位，电离势
penetration depth 穿透深度
coherence length 相干长度，相干距离
latent heat 潜热，潜伏热
second order phase transition 二级相变
specific heat 比热

Phrases

no sign of 没有征兆，没有特征
correlate with 找出一一对应的关系，相互关联影响
derive from 源出，来自

Notes

(1) Associated with the superconductive state is a perfect diamagnetism in which the magnetic flux is expelled from the material. This is known as the Meissner effect. This magnetic property associated with superconductors cannot be accounted for by the assumption that the superconducting state is equivalent to zero electrical resistance.

in which 引导定语从句；account for 译为“对……负责，说明……的原因”；that 引导宾语从句。

参考译文：与超导状态相关的是一种完全的抗磁性，其中磁通量从材料中排出。这就被称为迈斯纳效应。这种与超导体相关的磁性性质不能用超导态等价于零电阻的假设来解释。

(2) An attractive interaction between electrons through electron-lattice-electron interaction can lead to a ground state separated from excited states by an energy gap.

a ground state 为化学中的“基态”；excited state 为“激发态”。

参考译文：通过电子-晶格-电子相互作用，电子之间的相互吸引作用可导致基态与激发态之间通过能量间隙分离。

(3) The criterion for the transition temperature, T_c, of an element is related to the electron density of orbital $D(\varepsilon_F)$ at the Fermi surface and the electron-lattice interaction U, which can be estimated from the electrical resistivity.

the criterion for 译为“……的标准，判据”；related to 译为“与……相关”。

参考译文：元素的转变温度 T_c 的判据与费米表面的轨道电子密度 $D(\varepsilon_F)$ 和电子晶格相互作用 U 有关，可以通过电阻率估计。

(4) Occurrence of superconductivity and the T_c associated with the superconducting elements have not been correlated with other physical attributes, such as atomic number, ionization potential, atomic orbital, crystal structure, etc.

correlated with 译为“符合，一致，与……对应”。

参考译文：超导的发生与超导元素相关的 T_c 与其他物理属性，如原子序数、电离势、原子轨道、晶体结构等，没有关联。

(5) Transition from normal conductivity to superconductivity is a virtual perfect second order phase transition; that is, there is no latent heat or a sharp finite discontinuity in the specific heat; therefore, it is a cooperative phenomenon.

that is 译为“也就是说”，后面引导宾语补足语；specific heat 译为“比热”。

参考译文：从正常电导率到超导的转变实际上是一个完美的二级相变；即比热不存在潜热或尖锐的有限不连续；因此，这是一种合作现象。

19.2 阅读材料

19.2 Reading Materials

Thermal Conductivity of Metals

Metals represent a vast number of materials that have been the backbone of industrial development during the past two centuries. The importance of this development upon future technological progress is undiminished and unquestionable. Whether in pure, elemental form or as new lightweight, high-strength alloys, metals are simply indispensable to modern industrial society. Metals are, of course, known for their lustrous and shiny appearance, for the irmalleability and ductility and, above all, for their ability to conduct electric current. Because of the ir myriad applications, it is highly desirable to be able to tailor the properties of metals to match and optimize their use for specific tasks. Often among the important criteria is how well a given metal or alloy conducts heat. The physical parameter that characterizes and quantifies the material's ability to conduct heat is called thermal conductivity, of ten designated by κ. Understanding the nature of heat conduction process in metals and being able to predict how well a particular alloy will conduct heat are issues of scientific and technological interest.

1. Carriers of Heat in Metals

We know that crystalline lattices support heat flow when an external thermal gradient is imposed on the structure. Thus, like every other solid material, metals possess a component of heat conduction associated with the vibrations of the lattice called lattice (or phonon) thermal conductivity, ρ. The unique feature of metals as far as their structure is concerned is the presence of charge carriers, specifically electrons. These point charge entities are responsible not just for the transport of charge (i. e., for the electric current) but also for the transport of heat. Their contribution to the thermal conductivity is referred to as the electronic thermal conductivity. In fact, in pure metals such as gold, silver, copper, and aluminum, the heat current associated with the flow of electrons by far exceeds a small contribution due to the flow of phonons, so, for all practical purposes the electronic term is less dominant, and one has to take into account the phonon contribution in order to properly assess the heat conducting potential of such materials.

In discussions of the heat transport in metals (and in semiconductors), one makes an implicit and essential assumption that the charge carriers and lattice vibrations (phonons) are independent entities. They are described by their respective unperturbed wave functions, and any kind of interaction between the charge carriers and lattice vibrations enters the theory subsequently in the form of transitions between the unperturbed states. This suggests that one can express the overall thermal conductivity of metals (and other solids) as consisting of two independent terms, the phonon contribution and the electronic contribution:

$$\kappa = \kappa_e + \kappa_p \tag{1}$$

These two electrons and phonons are certainly the main heat carrying entities. However, there are other possible excitations in the structure of metals, such as spin waves. For the most part, we shall not consider these small and often conjectured contributions.

It is important to recognize that the theory of heat conduction, whether in nonmetallic or metallic systems, represents an exceptionally complex many-body quantum statistical problem. As such, it is unreasonable to assume that the theory will be able to describe all the nuances in the heat conduction behavior of any given material or that it will predict a value for the thermal conductivity that exactly matches that obtained from experiment. What one really hopes for is to capture the general trend in the behavior of the thermal conductivity, be it among the group of materials or in regard to their temperature dependence and have some reasonably reliable guidelines and perhaps predictive power as to whether a particular class of solids is likely to be useful in applications where heat carrying ability is an important concern or consideration.

Because of the obvious practical relevance of noble metals such as copper, gold and silver and of various alloys containing copper, there has always been a strong interest in comparing the current and heat conducting characteristics of these materials. It is therefore no surprise that certain important empirical relations were discovered nearly 50 years before the concept of an electron was firmly established and some 80 years before the notion of band structure emerged from the quantum theory of solids. Such is the case of the Wiedemann-Franz law, which states that, at a given temperature, the thermal conductivity of a reasonably pure metal is directly related to its electrical conductivity; in other words, by making a simple measurement of electrical resistivity (conductivity σ equals inverse resistivity $1/\rho$, we essentially know how good such a metal will be as a heat conductor.

$$\frac{\kappa}{\sigma} = \kappa\rho = LT \tag{2}$$

In many respects, the Wiedemann-Franz law and the constant L that relates the thermal conductivity of pure metals to their electrical conductivity, called the Lorenz number, are the fundamental tenets in the theory of heat conduction in metals. Most of our discussion will focus on the conditions under which this law is valid and on deviations that might arise. Such knowledge is of great interest: on one hand it allows a fairly accurate estimate of the electronic thermal conductivity without actually performing the rather tedious thermal conductivity measurements; on the other hand, deviations from the Wiedemann-Franz law inform on the presence and strength of particular (inelastic) scattering processes that might influence the carrier dynamics.

2. The Drude Model

The first important attempt in the theoretical understanding of transport processes in metals is due to Paul Drude. The classical free-electron model of metals developed by Drude builds on the existence of electrons as freely moving noninteracting particles that navigate through a positively charged background formed by much heavier and immobile particles. In any case, the classical electron gas is then described using the language of kinetic theory. The straight-line thermal motion of electrons is interrupted by collisions with the lattice ions. Collisions are instantaneous events in which electrons abruptly change their velocity and completely "forget" what their direction of motion was just prior to the collision. Moreover, it is assumed that thermalization occurs only in the process of collisions, i. e. , following a collision, although having a completely random direction of its velocity, the speed of the electron corresponds to the temperature of the region where the collision occurred.

In defining conductivities, it is advantageous first to introduce the respective current densities. Since electrons carry both charge and energy, their flow implies both electric and heat currents. The term electric current density, designated J_e, is understood to represent the mean electric charge crossing a unit area perpendicular to the direction of flow per unit time. Let us assume a current flows in a wire of cross-sectional area A as a result of voltage applied along the wire. If n is the number of electrons per unit volume and all move with the same average velocity v, then in time dt the charge crossing the area A is $-nevA\mathrm{d}t$. Hence, the current density is:

$$J_e = -ne\bar{v} \tag{3}$$

In the absence of driving forces (electric field or thermal gradient), all directions of electron velocity are equally likely and the average velocity is zero; hence there is no flow of charge or energy.

Similar to the electric current density, one can define the thermal current density J_Q as a vector parallel to the direction of heat flow with a magnitude equal to the mean thermal energy per unit time that crosses a unit area perpendicular to the flow. Because the speed of an electron relates to the temperature of the place where the electron suffered the most recent collision, the hotter the place of collision the more energetic the electron. Thus, electrons arriving at a given point from the hotter region of the sample will have higher energy than those arriving at the same point from the lower-temperature region. Hence, under the influence of the thermal gradient there will develop a net flux of energy from the higher-temperature end to the lower-temperature side. If we take $n/2$ as both the number of electrons per unit volume arriving from the higher-temperature side and the same density of electrons arriving from the lower-temperature region, and assuming that the thermal gradient is small (i. e., the temperature change over a distance equal to the collision length is negligible), it is easy to show that the kinetic theory yields for the heat current density the expression:

$$J_Q = \frac{1}{3} l_e \bar{v} c_v (-\nabla T) = -\kappa_e \nabla T \tag{4}$$

Here l_e is the mean free path of electrons, c_v is their electronic specific heat per unit volume, and e is the thermal conductivity of electrons. Equation (4) is a statement of the Fourier law with the electronic thermal conductivity given as:

$$\kappa_e = \frac{1}{3} l_e \bar{v} c_v \tag{5}$$

The fact that we talk about the Drude model some 100 years after its inception indicates that, in spite of its shortcomings, the model provided convenient and compact expressions for electrical and thermal conductivities of metals and, with properly calculated parameters, offered a useful measure of comparison between important transport properties.

(Source: Terry M Tritt. *Thermal Conductivity—Theory, Properties, and Applications*. Berlin: Spring, 2004)

Words and Expressions

malleability *n.* 顺从，可锻性，展延性
carrier *n.* 载流子，载体
phonon *n.* 声子
inverse *a.* 相反的，倒转的，颠倒的
tenet *n.* 原理，原则，信条
deviation *n.* 差异，偏差
heat conduction ［热］热传导
lattice vibration ［晶］晶格振动，点阵振动
band structure ［物］能带结构

19.3　问题与讨论

19.3　Questions and Discussions

(1) Describe the phenomenon of superconductivity.

(2) Which materials possess the possibility of superconductivity?

(3) What is Meissner effect?

(4) The basic principles and paradox of BCS theory.

(5) Is superconductivity related to normal conductivity?

(6) Does magnetism and superconducting appear to be mutually exclusive?

单元 20 金属材料腐蚀
Unit 20 Corrosion of Metallic Materials

扫码查看
讲课视频

20.1 教学内容
20.1 Teaching Materials

All structural metals corrode to some extent in material environments. Bronzes, brasses, stainless steels, zinc, and aluminum corrode so slowly under the service conditions in which they are placed that they are expected to survive for long periods without protection. When these same metals are placed into contact with more aggressive corrodents, they suffer attack and are degraded. Ordinarily iron and steel corrode in the presence of both oxygen and water. If either of these ingredients are absent, corrosion will not take place. Conversely, corrosion is retarded by protective layers (films) consisting of corrosion products or absorbed oxygen.

Although other forms of attack must be considered in special circumstances, uniform attack is one form most common confronting the user of metals and alloys. Corrosion is the destructive attack of a metal by a chemical or electrochemical reaction. Deterioration by physical causes is not called corrosion, but is describe as erosion, galling, or wear. In some instances, corrosion may accompany physical deterioration and is described by such terms as erosion corrosion, corrosive wear, or fretting corrosion. The term Rusting applies to the corrosion of iron-based alloys with the formation of corrosion products consisting largely of hydrous ferric oxides. Nonferrous metals and alloys corrode, but do not rust.

There are nine basic forms of corrosion that metallic materials may be subject to:

(1) Uniform corrosion. A metal resists corrosion by forming a passive film on the surface. This film is formed naturally when the metal is exposed to air for a period of time. It can also be formed more quickly by a chemical treatment. For example, nitric acid if applied to an austenitic stainless steel will form this protective film. Such a film is actually a form of corrosion, but once formed it prevents future degradation of the metal, as long as the film remains intact.

(2) Intergranular corrosion. Intergranular corrosion is a localized form of corrosion taking place at the grain boundaries of a metal with little or no attack on the grain boundaries themselves. This results in a loss of strength and ductility. The attack is often rapid, penetrating deeply into the metal and causing failure.

(3) Galvanic corrosion. When two different metallic materials are electrically connected and placed in a conductive solution (electrolyte), an electric potential exists. This potential difference will provide a stronger driving force for the dissolution of the less noble (more electrically negative)

material. It will also reduce the tendency for the more noble material to dissolve. The precious metals of gold and platinum are at the higher potential (more noble, or cathodic) end of the series (protected end), while zinc and magnesium are at the lower potential (less noble, or anodic) end. It is this principle that forms the scientific basis for using such materials as zinc to sacrificially protect a stainless steel drive shaft on a pleasure boat.

(4) Crevice corrosion. Crevice corrosion is a localized type of corrosion occurring within or adjacent to narrow gaps or openings formed by metal-to-metal or metal-to-nonmetal contact. It results from local differences in oxygen concentrations, associated deposits on the metal surface, gaskets, lap joints, or crevices under a bolt or around rivet heads where small amounts of liquid can collect and become stagnant.

(5) Pitting. Pitting is a form of localized corrosion. Pitting may result in the perforation of water pipe, making it unusable even though a relatively small percentage of the total metal has been lost due to rusting. Pitting can also cause structural failure from localized weakening effects even though there is considerable sound material remaining.

(6) Erosion corrosion. Erosion corrosion results from the movement of a corrodent over the surface of a metal. The movement is associated with the mechanical wear. The increase in localized corrosion resulting from the erosion process is usually related to the removal or damage of the protective film. The mechanism is usually identified by localized corrosion.

(7) Stress corrosion cracking. Stress corrosion cracking occurs at points of stress. Usually the metal or alloy is virtually free of corrosion over most of its surface, yet fine cracks penetrate through the surface at the points of stress. Depending on the alloy system and corrodent combination, the cracking can be intergranular or transgranular. The rate of propagation can vary greatly and is affected by stress levels, temperature, and concentration of the corrodent.

(8) Biological corrosion. Corrosive conditions can be developed by living microorganisms as a result of their influence on anodic and cathodic reactions. This metabolic activity can directly or indirectly cause deterioration of a metal by the corrosion process.

(9) Selective leaching. When one element in a solid alloy is removed by corrosion, the process is known as selective leaching, dealloying, or dezincification. The most common example is the removal of zinc from brass alloys. When the zinc corrodes preferentially, a porous residue of copper and corrosion products remain. The corroded part often retains its original shape and may appear undamaged except for surface tarnish. However, its tensile strength and particularly its ductility have been seriously reduced.

(Source: Philip A, et al. *METALLIC MATERIALS: Physical, Mechanical, and Corrosion Properties*. New York: Marcel Dekeer Inc, 2003)

Words and Expressions

corrodent　*a.* 腐蚀的；*n.* 腐蚀剂
deterioration　*n.* 恶化，变坏
galling　*a.* 擦伤人的；*n.* 擦伤，磨损
rust　*v.* （使）生锈；*n.* 锈，铁锈
intergranular　*a.* （颗）粒间的，晶界的
negative　*a.* 否定的，负电的，阴极的

noble *a.* 贵族的，惰性的，稀有的
gasket *n.* [机] 垫圈，衬垫
bolt *n.* (门或窗的) 金属插销，螺栓
perforation *n.* 穿孔，贯穿
transgranular *a.* 穿晶的，晶内的
microorganism *n.* 微生物
leaching *n.* [冶] 浸出，[化] 浸析
anodic *a.* 阳极的
cathodic *a.* 阴（负）极的
dealloying *n.* 脱合金腐蚀，去合金化
dezincification *n.* 脱锌
tarnish *v.* 使失去光泽；*n.* 暗锈,无光泽
uniform attack 均匀腐蚀，全面腐蚀
corrosive wear 腐蚀磨损
fretting corrosion 摩擦磨蚀，接触腐蚀
nitric acid 硝酸
uniform corrosion 均匀腐蚀，全面腐蚀
galvanic corrosion [化] 电偶腐蚀，
potential difference 电势差
crevice corrosion 缝隙腐蚀，裂隙腐蚀
lap joint [机] 搭接接头，接搭处
rivet heads [机] 铆钉头
pitting corrosion 点蚀
erosion corrosion 冲刷腐蚀，磨损腐蚀
stress corrosion 应力腐蚀
biological corrosion 生物腐蚀
metabolic activity 新陈代谢活动
chemical corrosion 化学腐蚀

Phrases

be expected to 预计，被期待，有望
in the presence of 在……面前；在……存在的前提下
in some instances 在某些情况下
be exposed to 曝光，暴露在
result from 起因于，由……造成
except for 除了……以外，要不是由于

Notes

(1) Although other forms of attack must be considered in special circumstances, uniform attack is one form most common confronting the user of metals and alloys.

uniform attack 译为“均匀腐蚀”，是腐蚀的一种类型；although 引导条件状语从句。

参考译文：虽然在特殊情况下必须考虑其他形式的腐蚀，但均匀腐蚀是金属和合金使用者最常遇到的一种形式。

(2) Nitric acid if applied to an austenitic stainless steel will form this protective film. Such a film is actually a form of corrosion, but once formed it prevents future degradation of the metal, as long as the film remains intact.

if 引导一条件状语；a form of 译为“一种类型，……的形式”；as long as 译为“只要”。

参考译文：如果硝酸应用于奥氏体不锈钢，将会形成一层保护膜。这种膜实际上是腐蚀的一种形式，一旦形成，它就可以防止金属的进一步降解，只要该膜保持完整即可。

(3) Intergranular corrosion is a localized form of corrosion taking place at the grain boundaries of a metal with little or no attack on the grain boundaries themselves. This results in a loss of strength and ductility.

intergranular corrosion 译为“晶间腐蚀”；grain boundaries 译为“晶界”；results in 译为“导致，产

生”。

参考译文：晶间腐蚀是发生在金属晶界的局部腐蚀形式，发生在金属的晶界处，几乎不腐蚀晶界本身。这将导致强度和延展性的损失。

(4) It is this principle that forms the scientific basis for using such materials as zinc to sacrificially protect a stainless steel drive shaft on a pleasure boat.

principle 译为“准则，实质，原理”；“it is+被强调部分+that”构成的强调句型。

参考译文：正是这一原理，形成了用锌等材料来牺牲保护游船上的不锈钢传动轴的科学依据。

(5) It results from local differences in oxygen concentrations, associated deposits on the metal surface, gaskets, lap joints, or crevices under a bolt or around rivet heads where small amounts of liquid can collect and become stagnant.

result from 译为“起因于，由……造成”；后面的“where”引导定语从句。

参考译文：这是由于氧气浓度的局部差异，在金属表面、垫圈、搭接接头或螺栓下方或铆钉头周围的缝隙沉积引起，这些地方会有少量液体聚集和凝结。

(6) Corrosive conditions can be developed by living microorganisms as a result of their influence on anodic and cathodic reactions. This metabolic activity can directly or indirectly cause deterioration of a metal by the corrosion process.

anodic(阳极的)和 cathodic(阴极的)互为反义词；directly(直接地)和 indirectly(间接地)互为反义词。

参考译文：由于活的微生物对阳极和阴极反应的影响，它们可以形成腐蚀条件。这种代谢活动可以直接或间接导致腐蚀过程中金属的劣化。

(7) When the zinc corrodes preferentially, a porous residue of copper and corrosion products remain. The corroded part often retains its original shape and may appear undamaged except for surface tarnish.

except for 译为“除了……以外”；when 引导时间状语从句。

参考译文：当锌优先腐蚀时，铜的多孔残留物和腐蚀产物仍然存在。被腐蚀的部分通常会保持原来的形状，除了表面变暗外看起来没有损坏。

20.2　阅读材料

20.2　Reading Materials

Fundamentals of Corrosion

1. Introduction

Perhaps the most striking feature of corrosion is the immense variety of conditions under which it occurs and the large number of forms in which it appears. Numerous handbooks of corrosion data have been compiled that list the corrosion effects of specific material/environment combinations; still, the data cover only a small fraction of the possible situations and only for specific values of, for example, the temperature and composition of the substances involved. To prevent corrosion, to interpret corrosion phenomena, or to predict the outcome of a corrosion situation for conditions other than those for which an exact description can be found, the engineer must be able to apply the knowledge of corrosion fundamentals. These fundamentals include the mechanisms of the various forms of corrosion, applicable thermodynamic conditions and kinetic laws, and the effects of the major variables. Even with all of the available generalized knowledge of the principles, corrosion is in most cases a very complex process in which the interactions among many different reactions, conditions, and synergistic effects must be carefully considered. All corrosion processes show some common features. Thermodynamic principles can be applied to determine which processes can occur and how strong the tendency is for the changes to take place. Kinetic laws then describe the rates of the reactions. There are, however, substantial differences in the fundamentals of corrosion in such environments as aqueous solutions, non-aqueous liquids, and gases that warrant a separate treatment in this Section.

2. Corrosion in Aqueous Solutions

Although atmospheric air is the most common environment, aqueous solutions, including natural waters, atmospheric moisture, and rain, as well as man-made solutions, are the environments most frequently associated with corrosion problems. Because of the ionic conductivity of the environment, corrosion is due to electrochemical reactions and is strongly affected by such factors as the electrode potential and acidity of the solution. As described in the article "Thermodynamics of Aqueous Corrosion", thermodynamic factors determine under what conditions the reactions are at an electrochemical equilibrium and, if there is a departure from equilibrium, in what directions the reactions can proceed and how strong the driving force is. The kinetic laws of the reactions are fundamentally related to the activation energies of the electrode processes, mass transport, and basic properties of the metal/environment interface, such as the resistance of the surface films.

The fundamental kinetics of aqueous corrosion have been thoroughly studied. The simultaneous occurrences of several electrochemical reactions responsible for corrosion have been analyzed on the basis of the mixed potential theory, which provides a general method of interpreting or predicting the corrosion potential and reaction rates. The actual corrosion rates are then strongly affected by the environmental and metallurgical variables. Special conditions exist in natural order and some industrial systems where biological organisms are present in the environment and attach themselves to the structure. Corrosion is expected by the presence of the organisms and the biological films they produce, as well as the products of their metabolism, as described in the Appendix "Biological Effects" to the aforementioned article on environmental variables.

3. Corrosion in Molten Salts and Liquid Metals

These are more narrow but important areas of corrosion in liquid environments. Both have been strongly associated with the nuclear industry, for which much of the research has been performed, but there are numerous nonnuclear applications as well. In molten-salt corrosion, described in the article "Fundamentals of High-Temperature Corrosion in Molten Salts", the mechanisms of deterioration are more varied than in aqueous corrosion, but there are many similarities and some interesting parallels, such as the use of the E-pO^{2-} diagrams similar to the E-pH (Pourbaix) diagrams in aqueous corrosion. Although the literature on molten-salt corrosion is substantial, relatively few fundamental thermodynamic and kinetic data are available.

Liquid-metal corrosion, discussed in the article "Fundamentals of High-Temperature Corrosion in Liquid Metals", is of great interest in the design of fast fission nuclear reactors as well as of future fusion reactors, but is also industrially important in other areas, such as metal recovery, heat pipes, and various special cooling designs. Liquid-metal corrosion differs fundamentally from aqueous and molten-salt corrosion in that the medium, except for impurities, is in anonionized state. The solubilities of the alloy components and their variation with temperature then play a dominant role in the process, and preferential dissolution is a major form of degradation. Mass transfer is another frequent consequence of the dissolution process. At the same time, the corrosion is strongly affected by the presence of nonmetallic impurities in both the alloys and the liquid metals.

4. Corrosion in Gases

In gaseous corrosion, the environment is nonconductive, and the ionic processes are restricted to the surface of the metal and the corrosion product layers (see the article "Fundamentals of Corrosion in Gases"). Because the reaction rates of industrial metals with common gases are low at room temperature, gaseous corrosion, generically called oxidation, is usually an industrial problem only at high temperatures when diffusion processes are dominant. Thermodynamic factors play the usual role of determining the driving force for the reactions, and free energy-temperature diagrams are commonly used to show the equilibria in simple systems, while equilibria in more complex environments as a function of compositional variables can be examined by using isothermal stability diagrams. In the mechanism and kinetics of oxidation, the oxide/metal volume ratio gives some guidance of the likelihood that a protective film will be formed, but the major role belongs to conductivity and transport processes, which are strongly affected by the impurities and defect structures of the compounds. Together with conditions of surface film stability, the transport processes determine the reaction rates that are described in general form by the several kinetic rate laws, such as linear, logarithmic, and parabolic.

The most obvious result of oxidation at high temperatures is the formation of oxide scale. The properties of the scales and development of stresses determine whether the scale provides a continuous oxidation protection. In some cases of oxidation of alloys, however, reactions occur within the metal structure in the form of internal oxidation. Like corrosion in liquids, selective or preferential oxidation is frequently observed in alloys containing components of substantially different thermodynamic stability.

5. Thermodynamics of Aqueous Corrosion

Corrosion of metals in aqueous environments is almost always electrochemical in nature. It occurs when two or more electrochemical reactions take place on a metal surface. As a result, some of the elements of the metal or alloy change from a metallic state into a nonmetallic state. The products of corrosion may be dissolved species or solid corrosion products; in either case, the energy of the system is lowered as the metal converts to a lower-energy form. Rusting of steel is the best known example of conversion of a metal (iron) into a nonmetallic corrosion product (rust). The change in the energy of the system is the driving force for the corrosion process and is a subject of thermodynamics. Thermodynamics examines and quantifies the tendency for corrosion and its partial processes to occur; it does not predict if the changes actually will occur and at what rate. Thermodynamics can predict, however, under what conditions the metal is stable and corrosion cannot occur.

The electrochemical reactions occur uniformly or nonuniformly on the surface of the metal, which is called an electrode. The ionically conducting liquid is called an electrolyte. As a result of the reaction, the electrode/electrolyte interface acquires a special structure, in which such factors as the separation of charges between electrons in the metal and ions in the solution, interaction of ions with water molecules, adsorption of ions on the electrode, and diffusion of species all play important roles. One of the important features of the electrified interface between the electrode and the electrolyte is the appearance of a potential difference across the double layer, which allows the definition of the electrode potential. The electrode potential becomes one of the most important parameters in both the thermodynamics and the kinetics of corrosion.

The electrode potentials are used in corrosion calculations and are measured both in the laboratory and in the field. In actual measurements, standard reference electrodes are extensively used to provide fixed reference points on the scale of relative potential values. The use of suitable reference electrodes and appropriate methods of measurement will be discussed in the section "Potential Measurements with Reference Electrodes". One of the most important steps in the science of electrochemical corrosion was the development of diagrams showing thermodynamic conditions as a function of electrode potential and concentration of hydrogen ions. These potential versus pH diagrams, often called Pourbaix diagrams, graphically express the thermodynamic relationships in metal/water systems and show at a glance the regions of the thermodynamic stability of the various phases that can exist in the system. Their construction and application in corrosion, as well as their limitations, will be discussed in the section "Potential Versus pH (Pourbaix) Diagrams".

(Source: Miroslav I Marek. *School of Materials Engineering*. ASM International Website)

Words and Expressions

moisture *n.* 湿气水分
parabolic *a.* 抛物线的；比喻的
electrolyte *n.* 电解质
acidity *n.* 酸味，酸性
molten-salt 熔盐
kinetic law 动力学定律
synergistic effect 协同作用
aqueous solution 水溶液
ionic conductivity 离子导电率
electrode potential ［物化］电极电位（电势）
free energy 自由能
reference electrode 参比电极

20.3　问题与讨论

20.3　Questions and Discussions

(1) Describe the common form of attack for metals and alloys.

(2) What kind of physical deteriorations will happen during corrosion?

(3) The principle and driving force of galvanic corrosion.

(4) Compare crevice corrosion, pitting, erosion corrosion.

(5) What is selective leaching?

单元21　腐蚀防护方法
Unit 21　Corrosion Protective Method

扫码查看
讲课视频

21.1　教学内容
21.1　Teaching Materials

It is important to recognise that, in a real corrosion situation, the corrosion cell and the electrochemical reaction will not always be readily identified. Protective methods are broadly subdivided into four categories.

1. Materials Selection and Design

Due consideration to the suitability of a material coupled with care in design can alleviate many corrosion problems. Amongst the most important factors to be borne in mind are:

(1) Materials (metals, alloys, nonmetallic materials)—which should be chosen to take into account, cost, availability and suitability to the environment in which the item is to be placed.

(2) Contact between metals of different standard electrode potentials—designs should avoid contacting between dissimilar metals where the kinetics of attack on one metal surface are enhanced by the presence of the second.

(3) Geometry—localised attack can be minimised by the avoidance of areas particularly susceptible to erosion or cavitation.

(4) Mechanical factors—excessive stress, internal or externally applied, should be avoided especially in metals known to be vulnerable to stress-corrosion cracking. This will reduce the instances of corrosion fatigue or fretting corrosion.

(5) Surface conditions—surface conditions that enhance susceptibility to localised attack should be avoided, e. g. roughened surfaces, broken films of metal or oxide and weld spatter.

(6) Electrochemical protection—where possible, designs should include provision for cathodic or anodic protection or for the application of protective coatings.

(7) Alloyings peciality alloys provide an excellent means of corrosion prevention for certain applications. In tidal zones, for example, the use of Ni-Cu alloys in the construction of jetties and offshore oil platforms has proved effective against corrosive attack by seawater.

2. Modification of the Electrolyte

(1) Removal of the aggressive species. Six possible strategies may be identified, summarised as follows:

1) Elimination of dissolved oxygen usually by evacuation, nitrogen saturation or by means of oxygen scavengers such as hydrazine.

2) Elimination of acidity by neutralisation, by addition of lime, for example.

3) Elimination of dissolved salts by means of reverse osmosis or ion exchange, for example.

4) Reduction of humidity by means of desiccants such as silica gel.

5) Reduction of local humidity by means of a localised temperature increase of around 5℃.

6) Elimination of solid particles in order to prevent deposit corrosion.

(2) Addition of corrosion inhibitors. A corrosion inhibitor is an inorganic or organic species that retards corrosion when introduced, in low concentration, into an aqueous solution in contact with the metal surface. Several different mechanisms exist for the inhibition.

1) Inhibitors may act by adsorption on the surface of the metal. Adsorption occurs around the corrosion potential and reduces the rate of either the anodic or cathodic reaction. Various sulphur, arsenic or phosphorus compounds act in this way.

2) An alternative mechanism for inhibition involves the formation of the precipitate on the metal surface or catalysis of a passivating reaction. Phosphonate, polyphosphate or hydrogen carbonate salts belong to this category of inhibitor.

3) A third group of inhibitors comprises redox reagents which act by shifting the surface potential to a region where cathodic or anodic protection occurs. The dichromate ion ($Cr_2O_7^{2-}$) acts by this type of mechanism.

3. Change of Electrode Potential

(1) Cathodic Protection

Cathodic protection may be achieved by means of a sacrificial anode or by the use of impressed current. Use of a sacrificial anode involves the construction of a galvanic cell between the metal substrate and the introduced anodes which are made of a different metal. Zinc, magnesium and certain aluminium alloys are commonly used as sacrificial anodes in the protection of steel. A steel ship's hull is often fitted with zinc blocks, for example, which are simply removed and replaced when necessary. In impressed current cathodic protection, a power supply is employed to drive current to the protected metal which serves as a cathode with respect to an auxiliary anode. The cathode potential is maintained within the required limits by current or cell voltage regulation in a case such as a pipeline or by a large scale potentiostat (controlled potential power supply) for an offshore construction.

(2) Anodic Protection

The principle of anodic protection depends on maintaining a stable passivating layer on the metal surface. Addition of elements such as palladium or copper as low concentration components in alloy steel produces galvanic anodic protection of the steel. Impressed current anodic protection is used to a much lesser extent than its cathodic counterpart. On commissioning, the impressed current must exceed the critical value necessary to passivate the surface. Thereafter, a reduction in current occurs such that the impressed current maintains the passive film. Stainless steel and titanium alloys, for example, may be protected in an acidic electrolyte by means of impressed current anodic protection.

4. Surface Coatings

A diverse range of surface coatings may be applied to protect a metal surface. Due consideration must be paid to the avoidance of localised damage that could render areas of the surface vulnerable to corrosion. Examples of surface coating protection layers include: Paint or a polymer cladding; Metal oxides, as in the anodising of aluminium; Metal coating, eg., steel sheet electroplated with zinc or hot-dipped galvanised iron. In all cases of protective coatings, the consequences of defects in the coating or damage of the coating for corrosion must be taken into account. Examples are shown where galvanic corrosion can result at pore sites or damaged sites in metal coatings and atmospheric corrosion may result at a damaged zone in a paint coating on a metal.

(Source: F Walsh, et al. *Corrosion and Protection of Metals.* Transaction of the Institute of Metal Finishing, 1993, 71: 117-120)

Words and Expressions

inhibitor　*n.* 抑制剂，缓蚀剂
catalysis　*n.* 催化
passivating　*n.* [化] 钝化，形成保护膜
jetty　*n.* 防波堤，码头
scavenger　*n.* 食腐动物，清道夫，清除剂
lime　*n.* 酸橙饮料，淡黄绿色，石灰
desiccant　*a.* 去湿的；*n.* 干燥剂
auxiliary　*a.* 辅助的，备用的
potentiostat　*n.* 稳压器，电位仪
osmosis　*n.* [物] 渗透，渗透性
cavitation　*n.* 气蚀，成洞，[流] 气穴现象
standard electrode potential　标准电极电势
dichromate ion　重铬酸盐离子
sacrificial anode　牺牲阳极
impressed current　外加电流，强制电流
anodic protection　阳极保护
galvanic cell　原电池，一次（自发）电池

Phrases

due consideration to　适当考虑，兼顾公平
take into account　考虑到，重视，体谅
on commissioning　现场调试，试运转

Notes

(1) Contact between metals of different standard electrode potentials-designs should avoid contacting between dissimilar metals where the kinetics of attack on one metal surface are enhanced by the presence of the second.

standard electrode potential 为电化学领域的“标准电极电势”；“where”引导定语从句，修饰前面的“metals”。

参考译文：不同标准电极电位金属之间的接触-设计时应避免不同金属之间的接触，因为第二种金属的存在会增强对第一种金属表面的侵蚀动力学。

(2) A corrosion inhibitor is an inorganic or organic species that retards corrosion when introduced, in low concentration, into an aqueous solution in contact with the metal surface.

corrosion inhibitor 译为“缓蚀剂”；“that”引导宾语从句，修饰“species”；“when”引导条件状语

从句。

参考译文：缓蚀剂是一种无机或有机物质，当低浓度引入到水溶液中时，与金属表面接触时可以延缓腐蚀。

(3) Use of a sacrificial anode involves the construction of a galvanic cell between the metal substrate and the introduced anodes which are made of a different metal. Zinc, magnesium and certain aluminium alloys are commonly used as sacrificial anodes in the protection of steel.

sacrificial anode 译为“牺牲阳极”，该法是电化学防腐的方法之一；“which”引导定语从句，修饰“anodes”。

参考译文：牺牲阳极法的使用是指在金属基底和引入由不同金属制成的阳极之间构造原电池。锌、镁和某些铝合金通常用作保护钢的牺牲阳极。

(4) In impressed current cathodic protection, a power supply is employed to drive current to the protected metal which serves as a cathode with respect to an auxiliary anode.

cathodic protection 译为“阴极保护”，是电化学防腐的另一种方法；“which”引导定语从句。

参考译文：在外加电流阴极保护中，采用电源将电流驱动到受保护的金属，该金属用作辅助阳极的阴极。

(5) On commissioning, the impressed current must exceed the critical value necessary to passivate the surface. Thereafter, a reduction in current occurs such that the impressed current maintains the passive film.

critical 此处为“临界”之意；such that 译为“如此……以至于”。

参考译文：在调试时，外加电流必须超过钝化表面所需的临界值。此后，电流减少，外加电流维持着该钝化膜。

21.2　阅读材料

21.2　Reading Materials

Methods for Corrosion Protection of Metals at the Nanoscale

1. Introduction

In 2016, the global cost of corrosion is estimated to be US $2.5 trillion (~3.4% of the global Gross Domestic Product, GDP). In case of conventional metallic structures, the acceptable value of thickness lost due to the metal degradation is about 100μm/year. Nowadays, with the development of nanoscience and nanotechnology, the small metallic parts have been widely used in many products, such as print electronics, interconnection, implant, nano-sensors, display units, ultrathin layers, drug delivery systems… Thus, their thickness loss should be controlled with acceptable values in range from 10 to 100nm.

Traditional methods for protection of metals include various techniques, such as coatings, inhibitors, electrochemical methods (anodic and cathodic protections), and metallurgical design. In practice, effective corrosion control is achieved by combining two or more of these methods. Usually, highly corrosion resistant materials are associated with a high cost factor. Even then, such materials can undergo degradation in severe environments/stress. The use of cheaper metallic materials along with proper corrosion control strategies is therefore economic for many applications. Nanomaterials and nanotechnology based protective methods can offer many advantages over their traditional counterparts, such as protection for early-stage, higher corrosion resistance, better corrosion control, and controlled release of corrosion inhibitors.

This mini review explores how metals can be protected at nanoscale by using both nanotechnology and nanomaterials (nano-alloys, nano-inhibitors, nano-coatings, nano-generators, nano-sensors).

2. Nano-Alloys

Nano-alloys (nanostructured alloys) are constructed from at least two different metallic nanomaterials, in order to overcome the limits of single components, to improve properties, to achieve new properties, and/or to achieve multiple functionalities for single metallic nanoparticles. Nano-alloys can also refer to the formation of nanocrystalline metal phases within the metallic matrices.

It was reported in literature that nano-alloys can offer many advantages over their conventional counterparts, such as higher corrosion resistance, high oxidation resistance, strong ductility enhancement, high hardness, and wear resistance. To improve the corrosion resistance of Ti-6Al-4V alloy without modifying its chemical composition, Kumar et al. fabricated the surface nanostructure for this alloy by using ultrasonic shot peening (USSP) method. Similarity used USSP method to fabricate the surface nanocrystallized AISI 409 stainless steel, for higher corrosion resistance. For high-strength aluminum alloys used the thermo-mechanical treatment to impart them the nanocrystalline structures and to control their strength and ductility prepared the 3D honeycomb nanostructure-encapsulated magnesium alloys. In their study, graphene oxide was incorporated into AZ61 alloy (at 1 wt%) to form the honeycomb nanostructure-encapsulated Mg alloys, which have the higher corrosion resistance and mechanical properties than the pure AZ61 alloy.

3. Nano-Inhibitors

For smart anticorrosion coatings, nano-inhibitors might refer to the inhibitor loaded nanocontainers. These nanocontainers exhibited the smart releasing property for their embedded inhibitors, by external or internal stimuli (such as pH-controlled release, ion-exchange control, redox-responsive control of release, light-responsive controlled-release, and release under mechanical rapture). In addition, the smart nanoshells could prevent the direct contact between the inhibitors with both coating matrices and adjacent local environments. All nanocontainers can be divided into two category of polymer nanocontainers (core-shell capsules, gels) and inorganic nanocontainers (porous inorganic materials). Available porous inorganic materials can be directly applied as inorganic nanocontainers for self-healing coatings. Inorganic nanocontainers could be mesoporous silica or titania, ion-exchange nanoclays and halloysite nanotubes. For organic coatings, various inhibitors have been loaded into nanocontainers, such as benzotriazole, mercaptobenzothiazole, mercaptobenzimidazole, hydroxyquinoline, dodecylamine, cerium salts, fluoride salt, zinc salts.

Nano-inhibitors can quickly respond to the local environmental changes associated with corrosion processes, such as local pH, ionic strength, and potential, and release encapsulated corrosion inhibitors to retard the corrosion process. Recently we reported the use of cerium load nanosilica for electroplating of Zn-Ni alloy coating on steel substrate. In our study, the electrochemical measurements suggested that inhibitor could be released from nanocontainer in the early electrodeposition of alloy coating on steel substrate. Whereas the salt spray test indicated that inhibitor was released during the corrosion of coated steel, thus increased the protective duration of coating by two times.

4. Nanocoatings

Nanocoatings (nanocomposite/nanostructured coatings) refer to the use nanomaterials and nanotechnology to enhance the coating performance. Nowadays coating should not only serve as the decoration with physical barrier, but also act as the smart multifunctional materials. For anticorrosion, in general, the barrier performance of coatings can be significantly improved by the incorporation of nanoparticles, that decreasing the porosity and zigzagging the diffusion path for deleterious species. At the interface coating/metal, nanoparticles are expected to enhance the coating adhesion and reduce the trend for the coating to blister or delaminate. Besides, nanoparticles or nanostructured coatings could act as a barrier against corrosive species.

It was reported in literature that nanomaterials could be used as nanofillers to reinforce both organic and metallic coatings. Various organic matrices have been used to fabricate the polymeric nanocomposite coatings, such as epoxy, polyurethane, chitosan, polyethylene glycol, polyaniline, rubber-modified polybenzoxazine, ethylene tetrafluoroethylene, polyester, polyacrylic, polydimethylsiloxane, polypyrrole, and alkyds. For metallic nanocoatings, two main matrices have been used for anticorrosion: Ni matrix and Zn-Ni matrix.

5. Nano-Generators

Nano-generators refer to the uses of nanosized devices/materials to convert the mechanical/thermal/light energies into electricity. It was reported in literature for the self-powered cathodic protection using nanogenerators.

Guo W. reported the use of the disk Triboelectric Nanogenerator (TENG) to provide a self-powered cathodic protection for stainless steels. Zhang H. reported the use of flexible hybrid nanogenerator (NG) for simultaneously harvesting thermal and mechanical energies. In their hybrid NG, a triboelectric NG was constructed below the pyro/piezoelectric NG. Recently Cui et al. reported the use of polyaniline nanofibers to construct a wind-driven TENG. Their TENG exhibited a high output values: maximum output voltage of 375V, short current circuit of 248μA, and 14.5mW power, under a wind speed of 15m/s.

In other direction, other approach of photo generated cathode protection, had been reported by coupling the nano-TiO_2 photo anode with metal electrode using simulated solar irradiation, white-light irradiation or under UV light. Recently, the hybridization of noble metals (Au, Ag, Pd) nanoparticles and nano-TiO_2 particles are the most promising approach not only to enhance the visible light sensitivity of TiO_2, but also to reduce the recombination of photo generated electron-hole pairs.

6. Nano-Sensors

Nano-sensors(or nanomaterials based sensors) can offer many advantages over their microcounterparts, such as lower power consumption, high sensitivity, lower concentration of analytes, smaller interaction distance between object and sensor. Besides, with the supports of artificial intelligence tools(such as fuzzy logic, genetic algorithms, neural networks, ambient-intelligence ⋯), sensor systems nowadays become smarter.

For corrosion protection at the nanoscale using smart coatings, the early detection of localized corrosion is very important, with regard to the economy, safety, and ecology. The most promising approach is to embed the smart nano-sensors, which are sensitive to the changes of environmental pH values, into the protective coating. Recently, Exbrayat developed the new nano-sensors for monitoring early stages of steel corrosion in NaCl solution. Their nano-sensors were constructed using silica nanocapsules, with the hydrophobic liquid core containing a fluorescent dye. In case of steel corrosion, these nano-sensors were able to detect iron ions and low pH values.

7. Conclusion and Future Scope

To protect metals from corrosion at the nanoscale, various methods could be used effectively, such as by using nano-alloys, nano-inhibitors, nano-coatings, nano-generators, nano-sensors ⋯ These advanced methods can not only protect metals at the early-stage, but also provide the structural health monitoring and self-healing. Besides the methods for corrosion protection at the nanoscale, there are several important studies should be carried out, such as(1) Mathematical modeling and simulation of corrosion at the nanoscale, (2) Methods for testing and measurement of metal corrosion at the nanoscale, (3) Nanoscale simulation of cathodic protection⋯

(Source: Mahdi Yeganeh, et al. *Methods for Corrosion Protection of Metals*. J. NANOSCI NANOTECHNO, 2019)

Words and Expressions

nanostructure *n.* 纳米结构
redox *n.* 氧化还原反应，氧化还原剂
benzotriazole *n.* 苯并三唑，连三氮杂茚
mercaptobenzothiazole *n.* 巯基苯并噻唑
mercaptobenzimidazole *n.* 巯基苯并恶唑
dodecylamine *n.* 十二烷胺
hydroxyquinoline *n.* 羟基喹啉
substrate *n.* 培养基，基质，底物
triboelectric *a.* 摩擦电的，摩电式的
sensitivity *n.* 灵敏度
hydrophobic *a.* 疏水的
drug delivery 药物传输，药物输送
graphene oxide 氧化石墨烯
core-shell capsule 核壳胶囊
power consumption 能量功耗
artificial intelligence 人工智能
fuzzy logic 模糊逻辑

21.3 问题与讨论
21.3 Questions and Discussions

(1) How to protect metals form corrosion?

(2) How to select and design materials?

(3) Two approaches of modifying electrolyte.

(4) The principle of cathodic protection and anodic protection.

(5) Examples of surface coating protection layers.

第六部分　金属材料的应用

Part 6　Applications of Metallic Materials

单元 22　多相金属催化基础
Unit 22　Fundamentals of Heterogeneous Metal Catalysis

扫码查看
讲课视频

22.1　教学内容
22.1　Teaching Materials

1. Introduction

Catalysis refers to the phenomenon by which the rate of a chemical reaction is accelerated by a substance(the catalyst) not appreciably consumed in the process. The term catalysis was coined by Berzelius in 1835 and scientifically defined by Ostwald in 1895, but applications based on catalysis can be traced back to thousands of years with the discovery of fermentation to produce wine and beer. Nowadays, catalysts are used in over 90% of all chemical industrial processes and contribute directly or indirectly to approximately 35% of the world's gross domestic product(GDP). Catalysis is central to a myriad of applications, including the manufacture of commodity, fine, specialty, petro-, and agrochemicals as well as the production of pharmaceuticals, cosmetics, foods, and polymers. Catalysis is also an important component in new processes for the generation of clean energy and in the protection of the environment both by abating environmental pollutants and by providing alternative cleaner chemical synthetic procedures.

Most catalytic processes are heterogeneous in nature, offering inherent advantages because of their ease of preparation, handling, separation from the reaction mixture, recovery, and reuse and also in terms of their stability, low cost, and low toxicity. Heterogeneous catalysts are most commonly made of metals and/or metal oxides, although metal sulfides, nitrides, carbides, phosphates and phosphides, ion-exchange resins, and clays are also employed in selected applications. Metal catalysts, especially those containing transition metals, have proven particularly useful in industrial catalysis because they have the ability to easily activate key molecules such as H_2, O_2, N_2, and CO as well as polyatomic organic molecules with C—H, C—O, C—N, and C-Cl bonds. Table 1 summarizes some of the most important metal-based catalytic processes developed since the 1870s.

Table 1　Important industrial heterogeneous catalytic processes promoted by metals

Process	Year	Main uses	Key reaction scheme	Typical catalyst
SO_2 oxidation to sulfuric acid	1875	Chemicals, metallurgic processing	$SO_2+\frac{1}{2}O_2 \rightarrow SO_3$	Pt on asbestos, $MgSO_2$ or SiO_2 (replaced by V_2O_5-K_2SO_4/SiO_2 since the 1920s)

Continued Table 1

Process	Year	Main uses	Key reaction scheme	Typical catalyst
Methanol to formaldehyde	1890	Resins for adhesives	$CH_3OH+\frac{1}{2}O_2 \rightarrow HCHO+H_2O$	Ag wire gauze or Ag crystals
Olefin hydrogenation	1902	Oil refining	$C_2H_4+H_2 \rightarrow C_2H_6$	Ni and Pt
Hydrogenation of edible fats and oils	1900s	Food production	Unsaturated fatty acids →partially saturated acids	Ni on a support
Methanation	1900s	Fuels	$CO+3H_2 \rightarrow CH_4+H_2O$	Ni on Al_2O_3 or other oxide supports
Ammonia synthesis (Haber)	1913	Fertilizers	$N_2+3H_2 \rightarrow 2NH_3$	Fe promoted with Al_2O_3、K_2O、CaO_2 and MgO
Ammonia oxidation (Ostwald)	1906	Nitric acid production	$4NH_3+5O_2 \rightarrow 4NO+6H_2O$	90%Pt-10% Rh wire gauze
Fisher-Tropsch synthesis	1938	Fuels	$CO+H_2 \rightarrow$ paraffins	Fe or Co with promoters on support
Steam reforming	1926	Synthesis gas	$C_mH_m+nH_2O \rightarrow nCO+[n+(m/2)]H_2$	Ni on support promoted by K_2O
Ethylene to ethylene oxide	1937	Antifreeze	$C_2H_4+\frac{1}{2}O_2 \rightarrow (CH_2)_2O$	Ag on α-Al_2O_3, promoted by Cl and Cs
Hydrogen cyanide synthesis	1930s	Chemicals	$CH_4+NH_3+\frac{3}{2}O_2 \rightarrow HCN+3H_2O$	90%Pt-10% Rh wire gauze
Catalytic reforming	1940s	Fuels	For example, n-$C_6H_{14} \rightarrow i$-C_6H_{14}	Pt, Pt-Re, or Pt-Sn on acidified Al_2O_3 or zeolite
Benzene to cyclohexane	1940s	Nylon	$C_6H_6+3H_2 \rightarrow C_6H_{12}$	Ni, Pt, or Pd catalysts
Vinyl acetate synthesis	1968	polymers	$C_2H_4+CH_3COOH+\frac{1}{2}O_2 \rightarrow CH_3COOCH{=}CH_2+H_2O$	Pd on SiO_2 or α-Al_2O_3
Automobile three-way catalysis	1970s	Pollution control	$CO+HC+NO_x+O \rightarrow CO_2+H_2O+N_2$	Pt, Pd, and Rh on monolith support

2. Fundamentals of Heterogeneous Metal Catalysis

(1) Kinetics Versus Thermodynamics, Active Centers, and Catalytic Cycles

Catalysis relies on changes in the kinetics of chemical reactions. Thermodynamics acts as an arrow to show the way to the most stable products, but kinetics defines the relative rates of the many competitive pathways available for the reactants, and can therefore be used to make metastable products from catalytic processes in a fast and selective way. Indeed, a catalyst can shorten the time needed to achieve thermodynamic equilibrium but cannot shift the position of that equilibrium, and therefore cannot catalyze a thermodynamically unfavorable reaction.

Figure 1 shows the microscopic picture of a heterogeneous catalytic reaction on the surface of a catalyst. These processes all start with the adsorption of the reactants. The surfaces of most catalysts are quite complex, though, and both the adsorption and the subsequent conversion reactions may take place preferentially at particular ensembles of surface sites, often called active sites or active centers. After diffusion toward and adsorption on the surface active sites, the reactants are converted to products (Figure 1). These products then desorb and diffuse out of the catalyst, leaving the active centers available for new incoming reactants. That way, the catalytic cycle can be repeated many times on each active site. For this to work, however, the bond strength between the adsorbed surface species and the active sites needs to be neither too weak nor too strong: too weak and the reactants cannot be readily activated; too strong and either the reactants are completely decomposed on the surface or the products cannot desorb.

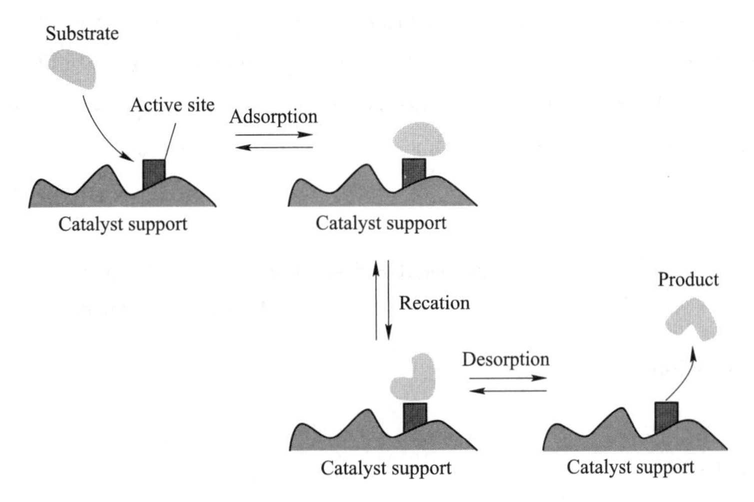

Figure 1 A scheme showing the main microscopic steps required in chemical conversions involving a catalyst surface

(2) Activity, Selectivity, and Stability

Three key parameters, activity, selectivity, and stability, can be used to evaluate catalyst performance. Achieving good and sustainable activity and selectivity at low cost is an everlasting goal in catalyst design and development. Activity refers to the rate at which a given reaction proceeds. In practice, activity can be expressed in absolute terms but is often reported using relative terms such as specific rate (rate divided by catalyst weight), areal rate (rate divided by surface area), turnover frequency (molecules converted per active center per unit time), conversion after specified time (fraction of reactants converted per fixed time), or required temperature for a given conversion under specific conditions.

In addition to activity, selectivity toward the desirable products also needs to be considered in the design of catalytic processes. In fact, selectivity is arguably the most important criterion to consider to decide on a particular catalytic process. Selective reactions consume less reactant, avoid the need of costly separations, and do not produce potentially polluting by-products.

Finally, there is the issue of stability. Ideally, catalysts are expected not to be consumed and to maintain a constant level of performance during the course of the reaction. In reality, however, both catalytic activity and catalytic selectivity may deteriorate during operation. The most common reasons for this catalyst deactivation include the evaporation, washout, reduction, corrosion, or other transformation of the active catalytic species during reaction, the formation of carbonaceous deposits, the poisoning due to strong adsorption of impurities such as sulfur or carbon monoxide on the surface of the catalyst, and/or the coalescence of metal particles (sintering).

(3) Supports and Promoters

As mentioned earlier, heterogeneous catalysis is centered around the chemistry of adsorbed species on solid surfaces. Consequently, metals are typically dispersed onto high-surface-area refractory supports such as alumina, silica, zeolites, activated carbon, titania, zirconia, or mixed oxides. The choice of support is primarily based on its surface area, but thermal and chemical stability, chemical properties, mechanical strength, and price also need to be considered.

To further optimize the performance of metal catalysts, promoters are often added. For instance, in ammonia synthesis, Al_2O_3, CaO, and MgO are added to the iron catalyst as textural promoters to minimize the sintering of Fe. Nowadays, almost all catalytic processes in industry involve the incorporation of certain promoters during catalyst preparation and/or the feed of additives with the reaction mixture.

(Source: R A Scott. *Heterogeneous Catalysis by Metal*. Encyclopedia of Inorganic and Bioinorganic Chemistry, 2014)

Words and Expressions

fermentation　*n.* 发酵生产
pollutant　*n.* 污染物
agrochemical　*n.* 农用化学品
toxicity　*n.* 有毒，毒性
phosphate　*n.* ［化］磷酸盐
phosphide　*n.* ［化］磷化物
adsorption　*n.* 吸附
desorb　*v.* 解吸，脱附，使释放
selectivity　*n.* 选择性
by-product　*n.* 副产物
deactivation　*n.* 减活作用，钝化作用
washout　*n.* 冲洗，冲刷
zeolite　*n.* 沸石
titania　*n.* ［化］二氧化钛，人造金红石
zirconia　*n.* ［化］氧化锆
promoter　*n.* 促进剂
ammonia　*n.* ［化］氨
ion-exchange resin　离子交换树脂
active site　活性中心
turnover frequency　交叉频率，转换频率
carbon monoxide　一氧化碳
activated carbon　活性炭

Phrases

trace back to　被追溯到
be converted to　被转换成
be expressed in　用……来表达

Notes

(1) Catalysis refers to the phenomenon by which the rate of a chemical reaction is accelerated

by a substance(the catalyst) not appreciably consumed in the process.

refer to 译为“涉及，参考，指的是”；by which 引导宾语从句。

参考译文：催化作用是指在反应过程中不被消耗（催化剂）而又加速化学反应速率的物质。

(2) Catalysis is also an important component in new processes for the generation of clean energy and in the protection of the environment both by abating environmental pollutants and by providing alternative cleaner chemical synthetic procedures.

both…and…连接两个并列句子。

参考译文：催化作用也是产生清洁能源新工艺和通过减少环境污染物，提供可替代的清洁化学合成工艺来保护环境的重要组成部分。

(3) Thermodynamics acts as an arrow to show the way to the most stable products, but kinetics defines the relative rates of the many competitive pathways available for the reactants, and can therefore be used to make metastable products from catalytic processes in a fast and selective way.

act as 译为“担当，起……作用，充当”；metastable 译为“亚稳的，介稳的”。

参考译文：热力学就像一个箭头，表示生成最稳定产物的方法，但动力学决定了反应物许多竞争途径的相对速率，因此可以快速地和选择性地从催化过程中生成亚稳产物。

(4) These products then desorb and diffuse out of the catalyst, leaving the active centers available for new incoming reactants. That way, the catalytic cycle can be repeated many times on each active site.

desorb（解吸）与 adsorb（吸附）互为反义词；active center 译为“活性中心”。

参考译文：这些产物随后脱附并扩散出催化剂，为新引入的反应物提供活性中心。这样，催化循环可以在每个活性部位重复多次。

(5) The most common reasons for this catalyst deactivation include the evaporation, washout, reduction, corrosion, or other transformation of the active catalytic species during reaction, the formation of carbonaceous deposits, the poisoning due to strong adsorption of impurities such as sulfur or carbon monoxide on the surface of the catalyst, and/or the coalescence of metal particles(sintering).

谓语 include 后面链接三个并列句子，是催化剂失活的原因。

参考译文：该催化剂失活的最常见原因包括反应过程中活性催化剂物种的蒸发、冲刷、还原、腐蚀或其他转化，碳质沉积的形成，由于催化剂表面强烈吸附硫或一氧化碳等杂质和/或金属颗粒的聚结（烧结）而中毒。

22.2 阅读材料

22.2 Reading Materials

Catalysis and the Adsorption of Hydrogen on Metal Catalysts

In any field of science as complex as that of catalysis, the only approach to its understanding is through a complete analysis of all factors involved, investigating each variable as completely separated from all other variables as possible. Heterogeneous catalysis always involves adsorption. In high temperature catalytic decompositions, the adsorption complex may be of very short duration. In bimolecular reaction the reactants must often both be adsorbed to yield the product; sometimes it is sufficient that one reactant be adsorbed. The rate determining step in heterogeneous reactions may be the rate of adsorption of the reactant on the surface or it may be the rate of desorption of the product from the surface. Only in rare cases is the rate of surface reaction proper determining.

By definition, the catalyst reduces the energy of activation of a reaction, and in heterogeneous catalysis this is achieved through the formation of an activated adsorption complex with the catalyst. This does not necessarily mean that strong adsorption is equivalent to great catalytic activity. In fact, too strong an adsorption may simply mean that the surface is covered with either reactant or product which effectively poisons the surface for any further reaction. This shows that the activation energy necessary to form the activated adsorption complex must be considered in relation to the activation energy necessary for the reaction proper within the activated adsorption complex. For instance, it has been conclusively shown that hydrogen must be adsorbed in the form of atoms in order to hydrogenate ethylene in the neighborhood of room temperature and atmospheric pressure. It has further been shown that ethylene itself is not adsorbed on the surface but that it merely picks up the two hydrogen atoms from the surface. This does not mean, of course, that ethylene, in the act of picking up the two hydrogen atoms, does not form momentarily an additional activated adsorption complex with the two hydrogen atoms and the surface. It only means that ethylene could not have been adsorbed itself prior to hydrogenation, since the sum of the separate heats of adsorption of hydrogen and ethylene is three times as great as the heat of hydrogenation. As may be readily concluded, adsorbed ethylene is a poison for this reaction. It was furthermore shown that the spacing of the hydrogen atoms on the surface that is, the crystal parameter of the surface-plays a very important part in the hydrogenation reaction, a 110 oriented nickel film, for instance, being five times as reactive as a randomly oriented film. The large spacing in rhodium of about 3.8Å ($1Å = 1\times10^{-10}$m) was found to be a thousand times more effective than the respective spacing in nickel of about 3.5Å. While this spacing may be of primary importance in the reaction at low temperatures, it is evident that at high temperatures-that is, with most of the molecules having a high energy content-the importance of the spacing should become less critical. While the required spacing for reactions at low temperatures calls for complete dissociation of the hydrogen molecule, complete dissociation is probably not necessary at high temperatures. Under conditions of incomplete dissociation, the heat of adsorption of hydrogen will necessarily be lower. Consequently, hydrogen will be adsorbed to a very much lesser degree on surfaces on which it is not completely dissociated, and in order to provide adequate coverage of the surface with adsorbed hydrogen molecules, very high pressures must be applied. But even in the regime of complete dissociation of the hydrogen molecules into adsorbed atoms, the heat of adsorption-that is, the binding energy of the atoms to the surface-must be considered in addition to spacing of the atoms. Atoms adsorbed with a high heat of adsorption may require (although not necessarily) a higher activation energy for the reaction between ethylene and adsorbed hydrogen atoms than those which are adsorbed with a low heat of adsorption. Hence it is of great importance to investigate whether the surface is uniform or nonuniform with respect to adsorption.

It is obvious that no truly uniform surfaces will exist. Even a perfect crystallographic plane is nonuniform due to periodic energy variations inherent in the atomic arrangement of the crystallographic sites. Next, one may ask whether all regions of highest surface energy that is, the crystallographic sites are equivalent. The answer is yes if the surface is a perfect crystallographic plane. If the plane is not perfect either through crystal imperfection or due to the incomplete atom layers which may form ledges and ridges, the crystallographic sites may vary in energy, but the maximum difference would be expected to be not more than 30% of the absolute value for a perfect surface, depending on the type of binding and on the size of the adsorbed molecule in relation to the size of the crystallographic sites. More serious is nonuniformity of the surface due to impurities. A foreign atom adsorbed on the surface with a large binding energy will make the surface nonuniform and will also decrease the surface energy of the neighboring sites. But the foreign atom need not necessarily be adsorbed on the surface; it may merely occupy a place in the crystallographic layer next to the surface. In this case too the resulting binding energy between the surface metal atoms and the foreign atom will lower the surface energy of the sites next to the foreign atom and thereby change the heat of adsorption for hydrogen atoms in this vicinity. Adsorbed hydrogen atoms themselves will have the same effect. They also will make the surface nonuniform. Beeck O found the heat of adsorption of hydrogen molecules to be 30000 calories per mole for a sparsely covered nickel surface. This means that the hydrogen atom is bound to the surface with an energy of roughly 65000 calories. For simplicity we may assume that this binding energy is equally distributed over the three or four neighboring metal atoms, depending on the type of the crystallographic site. It is obvious, therefore, that if a free site is surrounded by several other sites occupied by adsorbed hydrogen atoms, the energy of adsorption remaining for the adsorption of a hydrogen atom onto this free site must be much lower than the energy available for hydrogen atoms to be adsorbed on sites whose neighbors are still free. It is obvious too that under these conditions the decrease of heat of adsorption with increasingly covered surface will initially change little with surface coverage, since assuming mobility of adatoms from site to site the atoms can take positions far enough apart so as not to interfere with each other.

It will be shown that the heat of adsorption of hydrogen on porous evaporated metal films is often typical of this behavior, and mobility of adatoms will have to be assumed notwithstanding the observed relatively high heats of adsorption and the very low activation energy for adsorption. In this case, adsorption will take place on every site, and the redistribution of, for instance, adsorbed hydrogen will take place as adatoms through migration from site to site. If the adsorption process necessitates a high heat of activation as, for instance, in the case of the adsorption of nitrogen on iron at elevated temperatures, the assumption of mobility is not necessary, in as much as nitrogen molecules will initially not be adsorbed except on the sites of highest surface energy. The English school under the late J. K. Roberts of Cambridge has been quite successful in explaining the decrease of heat of adsorption with surface coverage of hydrogen on tungsten by the interaction of adatoms just discussed. Recently, Halsey and Taylor had rightly pointed out, however, that this interpretation must be limited to cases where the curves representing heats of adsorption vs. surface coverage intersect the ordinate with nearly right angles, the limiting value being an initial slope (change of heat of adsorption with surface coverage) not greater than about 0.25 the initial heat of adsorption. For larger angles heterogeneity of the surface will have to be assumed.

It is the purpose of this article to point out the need for caution in the interpretation of many experimental results so far published in the field of catalysis and for a critical attitude toward the problem of sorption of gases, which may include adsorption on the surface as well as absorption into the interior of the structure, and may easily lead to faulty conclusions.

(Source: Beeck O. *Catalysis and the Adsorption of Hydrogen on Metal Catalysts.* Advances in Catalysis, 1950: 151-195)

Words and Expressions

reactant *n.* 反应物，反应剂
hydrogenation *n.* ［化］加氢，氢化作用
dissociation *n.* 分解，分离，分裂
mobility *n.* 迁移率
adatom *n.* 吸附原子
redistribution *n.* 重新分布，重新分配
bimolecular reaction 双分子反应
randomly oriented 随机取向的

22.3 问题与讨论

22.3 Questions and Discussions

(1) What is the phenomenon of catalysis?
(2) The applications of catalysis.
(3) What is heterogeneous catalysis?
(4) What plays the important role in catalysis?
(5) The kinetics and thermodynamics of catalysis.
(6) How to evaluate catalyst performance?
(7) The principle of choosing supports.

单元 23　不锈钢在建筑行业中的应用
Unit 23　Applications of Stainless Steel in the Construction Industry

23.1　教学内容
23.1　Teaching Materials

扫码查看
讲课视频

1. Suitable Stainless Steels for Structural Applications

(1) Austenitic Stainless Steels

Austenitic stainless steels are generally selected for structural applications which require a combination of good strength, corrosion resistance, formability, excellent field and shop weldability and, for seismic applications, very good elongation prior to fracture.

Grades 1.4301 (widely known as 304) and 1.4307 (304L) are the most commonly used standard austenitic stainless steels and contain 17.5% to 20% chromium and 8% to 11% nickel. They are suitable for rural, urban and light industrial sites. Grades 1.4401 (316) and 1.4404 (316L) contain about 16% to 18% chromium, 10% to 14% nickel and the addition of 2% to 3% molybdenum, which improves corrosion resistance. They will perform well in marine and industrial sites. Grade 1.4318 is a low carbon, high nitrogen stainless steel which work hardens very rapidly when cold worked. It has a long track record of satisfactory performance in the railcar industry and is equally suitable for automotive, aircraft and architectural applications. Grade 1.4318 has similar corrosion resistance to 1.4301 and is most suitable for applications requiring higher strength than 1.4301 where large volumes are required. High chromium grades, containing about 20% chromium, are now available and will be introduced into EN 10088 in future revisions. Grade 1.4420 is an example of a high chromium (and high nitrogen) grade which has a claimed corrosion resistance similar to 1.4401. It is stronger than the standard austenitic grades, with a design strength of around $390N/mm^2$ compared to $240N/mm^2$, whilst retaining good ductility.

(2) Duplex Stainless Steels

Duplex stainless steels are appropriate where high strength, corrosion resistance and/or higher levels of crevice and stress corrosion cracking resistance are required. 1.4462 is an extremely corrosion resistant duplex grade, suitable for use in marine and other aggressive environments. An increasing use of stainless steels for load-bearing applications has led to increasing demand for duplex steels and development of new "lean" duplex grades. These grades are described as lean due to the reduced alloy contents of nickel and molybdenum which makes the grades significantly more cost effective. Lean grades have comparable mechanical properties to 1.4462 and a corrosion resistance which is comparable to the standard austenitic grades. This makes them appropriate for use in many onshore exposure conditions. Four lean duplex grades were added into EN 1993-1-4 in the 2015 amendment as they have become more widely available.

(3) Ferritic Stainless Steels

The two "standard" ferritic grades which are suitable for structural applications and commonly available are 1.4003 (a basic ferritic grade containing about 11% chromium) and 1.4016 (containing about 16.5% chromium, with greater resistance to corrosion than 1.4003). Welding impairs the corrosion resistance and toughness of grade 1.4016 substantially. Modern stabilised ferritic grades, for example 1.4509 and 1.4521, contain additional alloying elements such as niobium and titanium which lead to significantly improved welding and forming characteristics. Grade 1.4521 contains 2% molybdenum which improves pitting and crevice corrosion resistance in chloride containing environments (it has similar pitting corrosion resistance to 1.4401). 1.4621 is a recently developed ferritic grade that contains around 20% chromium, with improved polishability compared to 1.4509 and 1.4521.

2. Applications of Stainless Steel in The Construction Industry

Stainless steels have been used in construction ever since they were invented over one hundred years ago. Stainless steel products are attractive and corrosion resistant with low maintenance requirements and have good strength, toughness and fatigue properties. Stainless steels can be fabricated using a range of engineering techniques and are fully recyclable at end-of-life. They are the material of choice for applications situated in aggressive environments including buildings and structures in coastal areas, exposed to deicing salts and in polluted locations.

The high ductility of stainless steel is a useful property where resistance to seismic loading is required since greater energy dissipation is possible.

Typical applications for austenitic and duplex grades include:

(1) Beams, columns, platforms and supports in processing plant for the water treatment, pulp and paper, nuclear, biomass, chemical, pharmaceutical, and food and beverage industries.

(2) Primary beams and columns, pins, barriers, railings, cable sheathing and expansion joints in bridges.

(3) Seawalls, piers and other coastal structures.

(4) Reinforcing bar in concrete structures.

(5) Curtain walling, roofing, canopies, tunnel lining.

(6) Support systems for curtain walling, masonry, tunnel lining etc.

(7) Security barriers, hand railing, street furniture.

(8) Fasteners and anchoring systems in wood, stone, masonry or rock.

(9) Structural members and fasteners in swimming pool buildings.

(10) Explosion- and impact- resistant structures such as security walls, gates and bollards.

(11) Fire and explosion resistant walls, cable ladders and walkways on offshore platforms.

Ferritic grades are used for cladding and roofing buildings, as well as for solar water heaters and potable water pipes. They are also used for indoor applications such as elevators and escalators. In the transportation sector, they are used for load-bearing members, such as tubular bus frames. Although currently they are not widely used for structural members in the construction industry, they have the potential for greater application for strong and moderately durable structural elements with attractive metallic surface. For composite structures where a long service life is required, or where the environmental conditions are moderately corrosive, ferritic decking may provide a more economically viable solution than galvanized decking which may struggle to retain adequate durability for periods greater than 25 years. In addition to composite floor systems, other potential applications where ferritic stainless steel is a suitable substitute for galvanized steel include permanent formwork, roof purlins and supports to services such as cable trays. They could also be used economically in semi-enclosed unheated environments (e. g. railways, grandstands, bicycle sheds) and in cladding support systems, windposts and for masonry supports.

(Source: Afshan S, et al. *Design Manual for Structural Stainless Steel*, 2017: 3-7)

Words and Expressions

seismic *a.* 地震的，影响深远的
load-bearing *n.* 荷重
lean *v.* 倾斜；*a.* 贫乏的；*n.* 倾斜
polishability *n.* [材] 抛光性
beam *n.* 梁，光线，(体操的) 平衡木
column *n.* 圆柱，纵队，(报纸、杂志上的) 栏
platform *n.* 平台，讲台，(火车站) 月台
support *n.* 支撑物，载体，托
deicing *v.* 除冰
dissipation *n.* 消散，消耗
pulp *n.* 浆状物，纸浆，果肉
pharmaceutical *a.* 制药的，配药的
beverage *n.* 饮料
pin *n.* 别针，徽章，钉，楔，销，栓
railing *n.* 栏杆，扶手
seawall *n.* 海塘，防波堤
pier *n.* 码头，桥墩
curtain *n.* 窗帘，门帘，帘状物
canopy *n.* 天篷，罩，帐篷
masonry *n.* 石工 (行业)，石造建筑
fastener *n.* 扣件，钮扣，按钮
bollard *n.* [水运] 系船柱
decking *n.* 平台木板，盖板，装饰
cladding *n.* 包层，电镀，喷镀
formwork *n.* 模壳，模浆
purlin *n.* [建] 檩，桁条
grandstand *n.* 正面看台
shed *n.* 棚屋，厂房，工棚
ferritic stainless steel 铁素体不锈钢
cable sheathing 电缆包皮
galvanized steel 镀锌钢

Phrases

prior to 优先于
be suitable for 适合于
struggle to 竞争，尽力做某事

Notes

(1) Austenitic stainless steels are generally selected for structural applications which require a

combination of good strength, corrosion resistance, formability, excellent field and shop weldability and, for seismic applications, very good elongation prior to fracture.

which 引导定语从句；a combination of 译为“综合的”。

参考译文：通常选择奥氏体不锈钢通常被用于结构用途，这些结构应用需要有综合的性能，包括高强度、耐腐蚀性、可成形性、车间可焊性以及（对于地震应用）断裂前具有非常好的伸长率。

(2) These grades are described as lean due to the reduced alloy contents of nickel and molybdenum which makes the grades significantly more cost effective.

lean 作为形容词有“瘦的，贫乏的，精简的，效率高的，含量少的”之意，因此 lean alloy 译为“低合金钢”，lean ore 译为“贫矿”。

参考译文：由于镍和钼的合金含量低，这些牌号被称为贫级，这使得这些等级的钢更加贵重。

(3) They are the material of choice for applications situated in aggressive environments including buildings and structures in coastal areas, exposed to deicing salts and in polluted locations.

situate in 译为“位于，处于”，expose to 译为“接触，暴露于”。此处为定语从句，相当于省略了“which is”。

参考译文：它们是适用于恶劣环境的首选材料，包括沿海地区暴露在除冰盐和污染地区的建筑物和结构。

(4) For composite structures where a long service life is required, or where the environmental conditions are moderately corrosive, ferritic decking may provide a more economically viable solution than galvanized decking which may struggle to retain adequate durability for periods greater than 25 years.

where 引导定语从句, which 引导定语从句，修饰 galvanized decking（镀锌板）；struggle to 译为“竞争，尽力做”。

参考译文：对于需要较长使用寿命的复合材料结构，或环境条件具有中等腐蚀性的复合材料结构，铁素体层板可能提供比镀锌层板更经济可行的解决方案，镀锌层板可能难以保持 25 年以上的耐久性。

23.2　阅读材料

23.2　Reading Materials

Common Metals in Construction

Metal walls have become more popular in both commercial and residential construction, and it is easy to see why. They offer a multitude of design capabilities as well as a sustainable barrier to weather. They can be used for exterior and interior design, and come in several varieties. While they are used more often in commercial office and manufacturing buildings, you can find unique uses of these panels in residential homes and condos. More and more buildings are using metal in their construction. No wonder. There are so many benefits.

(1) Beauty. Metal Panels have a beautiful appearance for both exterior and interior applications. They have both shiny and matte finishes, and can be shaped and formed into a wide array of patterns. They can be made to look ultra-modern or vintage. A wide variety of colors, textures and designs makes them work with most commercial or residential architectural designs.

(2) Fireproof. Fire is always a risk in any building construction. Metal walls can act as a barrier to fire to help keep an active fire from spreading, and they also help keep the area cool in the case of too much heat.

(3) Rain Screen. Metal Panels also act as a rain screen and water barrier, preventing water from entering a building from the outside, and from spreading from room to room in the interior. In climates that get a lot of rain, having a metal barrier can extend the life of the building by preventing mold and decay from destroying other building materials.

(4) Lower Energy Consumption. Because of the density of metal, it offers a reduction in energy consumption that lowers expenses for building owners. Metal walls are considered to be eco-friendly and sustainable building materials, and meet building regulations for sustainable materials. Plus by using this construction, you are lowering your carbon footprint and the stress on the earth's resources.

(5) Durability. Metal walls are extremely durable, and hold up well to regular wear and tear. Their hardness may vary a bit depending on the metal used. You might be surprised that Metal Wall Panels do not dent or ding easily. They maintain their beauty for years.

(6) Low Installation and Maintenance Costs. Metal Wall Panels also save money on installation and maintenance. Panels can generally be installed more quickly than other materials such as brick, granite or precast. They are lighter in weight and often do not need much support structure as heavier materials.

1. Common Metals in the Construction Industry

Chosen for their durability, strength and resistance to weather, metals used in the construction industry serve a wide range of functions. The most common of them are carbon steel, aluminum, copper tubing and stainless steel, which each have their particular qualities and ideal uses. As a whole, however, these metals are ubiquitous in the world of buildings and architecture, in applications both small and large.

Carbon steel is one alloy that is prized in the construction industry for its hardness and strength. It is typically used to make beams for structural framework, plates for highway construction, and rectangular tubing for welded frames trailer beds, and bridges. It is also a material of choice to make rebar and hollow structural sections (HSS). Made by mixing carbon and iron together, carbon steel is classified on a scale of "mild" to "very high," depending on how much carbon is present in the metal.

Aluminum is also commonly used in the industry because it is resistant to corrosion, highly conductive and ductile. Because it is resistant to harsh weather, the metal is used in windows, doors, and wire, as well as outdoor signage and street lights. The metal is processed into sheets, tubes and castings, and also used to build automobiles and trucks, as well as bicycles and marine vessels. HVAC ducts, roofs, walling and handles made of aluminum are also frequently found in the building industry.

Copper tubing, which comes in two main types, is often used to construct pipes in buildings. Rigid copper tubing is ideal for hot and cold tap water pipes in buildings. Soft copper, on the other hand, is frequently used to make refrigerant lines in HVAC systems and heat pumps. Copper ductile, malleable metal is resistant to corrosion from water and soil, and is also recyclable. Copper tubing is also easily soldered, forming lasting bonds. All of these properties make this metal ideal for piping and tubing.

Stainless steel is among one of the oldest known building materials. It was used centuries ago to construct structures that still stand today, thanks to the corrosion and stain resistant properties of the metal. Some of the most famous architectural structures, such as the Chrysler Building in New York City, rely on stainless steel for its strength, durability and reliability. Stainless steel is an alloy of several different metals, the amounts of which can be adjusted to create different grades of stainless steel with different properties. The most common grade is 301, which is ductile and easily welded. It can be found in roofing, structural applications, handrails and balustrading, architectural cladding, as well as in drainage components.

2. Is Steel Still the Best Material for Building?

Steel has a long history in the construction industry, but is it still the best material for building?

Ever since the first skyscrapers went up in Chicago during the late 1800s, steel has been a major component in commercial building construction. Before that, builders used cast iron. But they found that structural steel beams set in concrete allowed them to frame tall buildings that were more fire resistant and more structurally sound than cast iron. Since that time, steel has not only become the best building material for commercial construction but closely tied to economic health. In fact, many experts look to the steel industry as an indicator of how well the economy is doing. Steel has a long history in the construction industry, but is it still the best material for building? The steel industry was not immune to the effects of the recent economic downturn. American steel producers like Butler Manufacturing have been facing layoffs, due to a slowdown in construction projects. Steel companies are trying to weather the economic storm just like other businesses, and less construction means fewer production jobs. Steel is also getting more expensive because of the price of raw materials for making steel, iron and coal, are on the rise. And while steel is still popular, other construction materials are giving it a run for its money.

3. New Building Materials

While no one alternative has become a standard to replace steel, materials like engineered timber and metal composites are becoming more common in new construction projects.

Timber companies tout wood as a durable, renewable resource, and engineered timber is gaining some traction as an alternative to steel. For example, the new arts and media building at Nelson Marlborough Institute of Technology in New Zealand used engineered wood in place of typical steel and concrete construction, and the company that worked on that building says that it's taking on more and more contracts that would have gone to steel construction companies.

Composite materials like Fiber Reinforced Plastics(FRP) and alternative metal alloys are gaining popularity in commercial construction, as well. Composites can be more durable than steel, and repairing damaged composite components is often less costly and requires less heavy machinery. The big drawback with these alternative materials right now is the cost. Because FRP and other composites are relatively new, they're still costlier to produce than steel components.

4. Residential Steel

In residential construction, steel is actually gaining popularity. In the past, builders preferred wood over steel for framing residential buildings, but its durability has some builders looking to steel as an alternative.

The major drawbacks to using steel in residential construction are price and energy use. Steel is becoming more common in residential buildings, but in many areas it is still hard to find contractors to build residential homes with steel framing. A 2002 U. S. Department of Housing study built a steel home alongside a wood home to compare the costs of the two materials. The steel home cost about 14 percent more to build and required more time to complete. However, steel has a higher strength to weight ratio than wood, meaning that steel components are stronger without adding much weight. That helps make steel structures stronger than wood, which is very attractive in areas prone to tornadoes, earthquakes, and other natural disasters. Steel is also fire- and termite-resistant, making it more durable than wood.

5. Steel's Pros and Cons

There are a couple of problems with using steel in construction. In very humid areas, coastal regions, or even in rooms like the bathroom that get very moist, steel will corrode unless builders use extra coatings of anti-corrosives to protect it. Also, since steel conducts heat and cold well, it's not ideal from an insulation standpoint.

Green builders use steel in eco-friendly construction projects because of its durability and renewability. Steel is long-lasting, and combined with other eco-friendly building materials is often used for green building projects. And unlike other recyclable materials such as plastic, steel doesn't lose quality each time it is recycled. There's also less waste associated with steel construction compared to wood, because you can weld small "offcuts" together to do smaller jobs. Despite a few drawbacks, steel is still the preferred material for framing commercial buildings and is gaining popularity for residential construction.

(Source: *Common Metals in the Construction Industry*. Continentalsteel Website)

Words and Expressions

fireproof　*a.* 防火的，耐火的
rebar　*n.* ［材］钢筋，螺纹钢（筋）
hollow　*a.* 中空的
duct　*n.* 输送管，导管，通风管道
refrigerant　*n.* 制冷剂
concrete　*n.* ［建］混凝土
drawback　*n.* 缺点，不利条件
timber　*n.* 木材
eco-friendly　*a.* 对生态环境友好的
offcut　*n.* （石、木、纸等的）边料

residential construction　住宅建设　　carbon footprint　碳足迹
maintenance cost　维护成本　　renewable resourse　可再生资源
support structure　支撑结构

23.3　问题与讨论

23.3　Questions and Discussions

(1) What kind of stainless steel is generally selected for structural applications and why?

(2) What are duplex stainless steels used for?

(3) What are the commonly available ferritic grades suitable for structural applications?

(4) Introduce typical applications for austenitic and duplex grades.

(5) The applications of ferritic SS in construction.

单元 24　铝合金在机身结构中的应用
Unit 24　Aluminum Alloys for Airframe Structures

扫码查看
讲课视频

24.1　教学内容
24.1　Teaching Materials

Aluminum alloys have been the dominant materials used for airframe structures until the increasing trend in the use of polymer matrix composites, as shown in Table 1. The Boeing 787 and Airbus A350 are built with about 50% of these materials, which are lighter than aluminum alloys and possess better resistance to degradation by corrosion and fatigue. The cost premium for composite structures is also offset by the lower operation and in-service maintenance costs compared with those for metallic structures and by continuous improvements in their manufacturing efficiency. In response, by focusing on an integrated approach to working with the users, the aluminum suppliers have continued to develop alloys with better mix of properties (Alcoa has developed 65 new specifications for aluminum alloys since 2000) and manufacturing characteristics, resulting in structures that are less expensive than those made from titanium and composites while also being up to 10% lighter at the system level. Although both the fuselage and wing skins of 777 and A350 principally used composite materials, the fuselage of the Boeing 777×, launched in 2013 and expected to enter into service in late 2019, will use metallic materials.

Table 1　Materials Usage in Commercial Airplanes

Material	Materials Distribuion (by weight)/%						
	Boeing					Airbus	
	747	757	767	777	787	A380	A350
Aluminum	81	78	80	70	20	61	19
Titanium	4	6	2	7	15	10(Ti and steel)	14
Steel	13	12	14	11	10		6
Composites	1	3	3	11	50	22	53
Other	1	1	1	1	5	7	8

The historical development of aluminum alloys for airframe structural applications has also been responsive to meeting the evolving criteria for airplane design, as shown in Table 2. The early effort was on developing alloys with as high a strength as possible, which, however, resulted in corresponding reductions of toughness and corrosion resistance. With the advent of fracture mechanics based design philosophy and the use of thicker product forms, durability, damage tolerance, and resistance to stress corrosion cracking (SCC), particularly in the short transverse direction, became important considerations for eliminating degradation by fatigue and corrosion and for reducing in-service maintenance costs. The overall goal of alloy development presently is to reduce the cost of manufacturing, operation, and maintenance of airplane systems without sacrificing property combinations and structural performance.

Table 2　Evolution of design drivers for commercial aircraft

Time frame	Principal design drivers
Pre—1930s	Static strength
1930s—1960s	Static strength, corrosion resistance, fatigue
1960s—1970s	Static strength, corrosion resistance, stable elevated temperature properties, durability, damage tolerance, good properties in thick sections
1980s	All the above and dramatic weight savings
1990s	Static strength, corrosion resistance, durability, damage tolerance, good properties in thick sections, highly balanced properties, low manufacturing, acquisition and maintenance (ownership) costs, high level of safety
2000s—present	Above as for 1990s and amenability to revolutionary processing such as additive manufacturing.
Future	Above plus breakthrough structural weight and cost savings, extended maintenance/inspection intervals, passenger comfort, green aerostructures

Although precipitation hardenable aluminum alloys were rapidly being developed and used, the basic alloy systems, Al-Cu-Mg and Al-Zn-Mg-Cu and the narrow choice of compositions on which they are based have remained the same to date. Despite this limitation, the improvements in properties enabling significant performance gains in airframes are due to the microstructural diversity that can be achieved by alloying and processing. This has been helped in no small measure by the overall development of the theories and understanding of the strengthening mechanisms in metals aided by the availability of continuously improving testing and characterization methods and more recently by computational modeling.

Only the precipitation hardened alloys based on the Al-Cu and the Al-Zn-Mg systems provide sufficient strength for use in airframes, with the former used for structures for which damage tolerance is critical and the latter where strength is critical. The tensile strength of pure, annealed aluminum is about 10MPa (~1.5ksi). All aluminum alloys contain iron and silicon as impurities, although silicon is also added deliberately to certain alloy families for strengthening. The yield strength of the commercially pure wrought aluminum alloy 1100 with a combined total maximum of 1 wt% Fe and Si is about 30MPa (~4ksi) in the annealed condition. The improvement in the yield strength of this alloy over that of pure aluminum can be attributed to very limited solid solution strengthening and wrought processing effects. In wrought aluminum alloys, most of the Fe and Si are present in the form of constituent particles in a size range greater than about 1mm that makes them ineffective for strengthening. The yield strength of alloy 1100 can be increased to about 150MPa (~20ksi) by work hardening. Alloys with Mn and Mg as the principal alloying elements can be work hardened to achieve yield strength levels of about 250MPa and 300MPa (~35 ~ 40ksi), respectively. Only precipitation hardening can provide higher strength levels in aluminum alloys.

As shown in Table 2, static strength is always the principal consideration for the selection of aluminum alloys. Tradeoffs between strength and other properties, for example fracture toughness, resistance to degradation by fatigue, and various types of corrosion are considered only to the extent that the alloys possess the minimum strength required for structural performance. Over the past decades, aluminum alloys have been developed with significant, concurrent improvements to both strength and other properties.

Precipitation hardening (solution heat treatment, quenching, and aging), which is key to achieving the strength, is just the last one prior to which many of the matrix microstructural features that affect the properties have already evolved. This scheme along with the ability to model the effects of composition, processing, and microstructure on properties will be used as a roadmap to gain an understanding of the underlying physical metallurgy of heat treatable aluminum alloys pertinent for alloy design.

(Source: Krishnank K Sankana, et al. *Metallurgy and Design of Alloys with Hierarchical Microstructures*. Amsterdam: Elsevier, 2017)

Words and Expressions

degradation　*n.* 毁坏，恶化（过程），降解
tradeoff　*n.* 权衡，折中
diversity　*n.* 多样性，多样化，差异
mitigating　*v.* 减轻，缓和
in-service　*a.* 在服役期间，在职的
aging　*n/v.* 老化，变老；［冶］时效处理
polymer matrix composite　聚合物复合材料
damage tolerance　损伤容限；损伤容忍度
stress corrosion cracking　［力］应力腐蚀开裂
strengthening mechanism　强化机理
static strength　静力强度，静态强度

Phrases

in response　作为响应，作为回答
up to　高达
be responsive to　配合为，回应，响应
along with　事物的结合；综合……
be attributed to　归因于……
pertinent for　与……有关/相关的
in no small measure　在很大程度上

Notes

(1) The cost premium for composite structures is also offset by the lower operation and in-service maintenance costs compared with those for metallic structures and by continuous improvements in their manufacturing efficiency.

in-service 译为“在职的，在服役期间的，使用中的”；compared with 译为“与……相比”；offset 译

为“抵消，补偿”。

参考译文：与金属结构相比，复合材料的可操作性和使用维护成本较低，以及其制造效率的不断提高，抵消了复合材料的成本优势。

(2) With the advent of fracture mechanics based design philosophy and the use of thicker product forms, durability, damage tolerance, and resistance to stress corrosion cracking (SCC), particularly in the short transverse direction, became important considerations for eliminating degradation by fatigue and corrosion and for reducing in-service maintenance costs.

该句的主语为 durability, damage tolerance, and resistance to stress corrosion cracking (SCC); damage tolerance 译为“损伤容限”。

参考译文：随着基于断裂力学的设计理念的出现以及更厚的产品形式的使用，耐久性、损伤容限和特别是在短的横向方向上的抗应力腐蚀开裂（SCC）成为消除疲劳和腐蚀退化和降低在役维护成本的重要考虑因素。

(3) This has been helped in no small measure by the overall development of the theories and understanding of the strengthening mechanisms in metals aided by the availability of continuously improving testing and characterization methods and more recently by computational modeling.

in no small measure 译为“在很大程度上”，aided by 译为“受助于”。

参考译文：这在很大程度上得益于理论的全面发展和对金属强化机制的理解，以及不断改进的测试表征方法和计算建模。

(4) Tradeoffs between strength and other properties, for example fracture toughness, resistance to degradation by fatigue, and various types of corrosion are considered only to the extent that the alloys possess the minimum strength required for structural performance.

tradeoff 译为“权衡，折中”；required for 译为“要求，需求”。

参考译文：强度和其他性能之间的权衡，例如断裂韧性、抗疲劳退化和各种类型的腐蚀，只有在合金满足所需结构性能的最小强度时才被考虑。

(5) Precipitation hardening (solution heat treatment, quenching, and aging), which is key to achieving the strength, is just the last one prior to which many of the matrix microstructural features that affect the properties have already evolved.

which 引导定语从句修饰 precipitation hardening（析出硬化，沉淀硬化）；to which 用来修饰 the last one；prior to 译为“在……之前，居先”；key to 译为“……的关键”。

参考译文：沉淀硬化（固溶热处理，淬火和时效处理）是获得强度的关键，而这些仅仅是最后一步，在此之前，许多影响基体性能的微观结构特征已经形成。

24.2 阅读材料

24.2 Reading Materials

Aerospace Materials —Past, Present, and Future

Forty years ago, aluminum dominated the aerospace industry. As the new kid on the block, it was considered to be lightweight, inexpensive, and state-of-the-art. In fact, as much as 70% of an aircraft was once made of aluminum. Other new materials such as composites and alloys were also used, including titanium, graphite, and fiberglass, but only in very small quantities. Readily available, aluminum was used everywhere from the fuselage to main engine components. Times have changed. A typical jet built today is as little as 20% pure aluminum. Most of the non-critical structural material-paneling and aesthetic interiors-now consist of even lighter-weight carbon fiber reinforced polymers(CFRPs) and honeycomb materials.

1. Aerospace Unique among Industries

Aerospace manufacturing is unique among other volume manufacturing sectors, and this is especially true of aerospace engine manufacturing. The engine is the most complex element of an aircraft, houses the most individual components, and ultimately determines fuel efficiency. The advent of lean-burn engines, with temperature potentials as high as 3800℉(2100℃), has helped drive demand for these new materials. Considering that the melting point of current super alloys is around 3360℉ (1850℃), the challenge becomes finding materials that will withstand hotter temperatures. To meet these temperature demands, heat-resistant super alloys(HRSAs), including titanium alloys, nickel alloys, and some nonmetal composite materials such as ceramics, are now being brought into the material equation. These materials tend to be more difficult to machine than traditional aluminum, historically meaning shorter tool life and less process security.

There's also a high process risk in machining aerospace parts. Because margins for error are non-existent at 35000ft cruising altitude, tolerances in aerospace are more precise than almost any other industry. This level of precision takes time. Longer machining times are required for each component, and more time per part makes scrap relatively expensive, when factoring in time investment. In addition, aerospace materials themselves impact component design. Design for manufacturability(DFM) is the engineering art of designing components with a balanced approach, taking into consideration both component function and its manufacturing requirements. This approach is being applied more and more in aerospace component design because its components have to accomplish certain loads and temperature resistances, and some materials can only accommodate so much. This give-and-take relationship between material and design is a particular consideration when investigating next-generation materials.

2. New Material Landscape

Standard aerospace aluminums—6061, 7050, and 7075 —and traditional aerospace metals—nickel 718, titanium 6Al4V, and stainless 15-5PH—still have applications in aerospace. These metals, however, are currently ceding territory to new alloys designed to improve cost and performance. To be clear, these new metals aren't always new, some having been available for decades. Rather, they are new to practical production application, as machine tools, tooling technology, and insert coatings have sufficiently advanced to tackle difficult-to-machine alloys.

Even though the amount of aluminum is declining in aircraft, its use is not completely disappearing. In fact, aluminum is coming back, especially in cases where the move to CFRP has been cost prohibitive or unsuccessful. Titanium aluminide(TiAl) and aluminum lithium(Al-Li), for example, which have been around since the 1970s, have only been gaining traction in aerospace since the turn of the century. Similar to nickel alloy in its heat-resisting properties, TiAl retains strength and corrosion resistance in temperatures up to 1112℉(600℃). But TiAl is more easily machined, exhibiting similar machinability characteristics to alpha-beta titanium, such as Ti_6Al_4V. Perhaps more importantly, TiAl has the potential to improve the thrust-to-weight ratio in aircraft engines because it's only half the weight of nickel alloys. Case in point, both low-pressure turbine blades and high-pressure compressor blades, traditionally made of dense Ni-based super alloys are now being machined from TiAl-based alloys. Another re-introduction of aluminum to aerospace is found in weight-saving Al-Li, specifically designed to improve properties of 7050 and 7075 aluminum. Overall, the addition of lithium strengthens aluminum at a lower density and weight, two catalysts of the aerospace material evolution. Al-Li alloys' high strength, low density, high stiffness, damage tolerance, corrosion resistance, and weld-friendly nature make it a better choice than traditional aluminums in commercial jetliner airframes. Airbus is currently using AA2050. Meanwhile, Alcoa is using AA2090 T83 and 2099 T8E67. The alloy can also be found in the fuel and oxidizer tanks in the SpaceX Falcon 9 launch vehicle, and is used extensively in NASA rocket and shuttle projects.

Titanium 5553(Ti-5553) is another metal that is reasonably new to aerospace, exhibiting high strength, light weight, and good corrosion resistance. Major structural components that need to be stronger and lighter than the previously used stainless steel alloys are perfect application points for this titanium alloy. Nicknamed triple 5-3, this has been a notoriously difficult material to machine-until recently. Some structural pieces, like fasteners, landing gear, and actuators, require raw strength, with lightweight properties being less of a priority. In such cases, Carpenter Technology Ferrium S53 steel alloy has provided mechanical properties equal to or better than conventional ultra-high-strength steels, such as 300M and SAE 4340, with the added benefit of general corrosion resistance. This can eliminate the need for cadmium coating and the subsequent related processing.

3. Composites Hit their Stride

Composite materials also represent a growing piece of the aerospace material pie. They reduce weight and increase fuel efficiency while being easy to handle, design, shape, and repair. Once only considered for light structural pieces or cabin components, composites' aerospace application range now reaches into true functional components—wing and fuselage skins, engines, and landing gear. Also important, composite components can be formed into complex shapes that, for metallic parts, would require machining and create joints. Preformed composite components aren't just lightweight and strong, they reduce the number of heavy fasteners and joints—which are potential failure points—within the aircraft. In doing so, composite materials are helping to drive an industry-wide trend of fewer components in overall assemblies, using one-piece designs wherever possible.

While CFRPs represent the lion's share of composite material in both cabin and functional components, and honeycomb materials provide effective and lightweight internal structural components, next-generation materials include ceramic-matrix composites (CMCs), which are emerging in practical use after decades of testing. CMCs are comprised of a ceramic matrix reinforced by a refractory fiber, such as silicon carbide(SiC) fiber. They offer low density/weight, high hardness, and most importantly, superior thermal and chemical resistance. Like CFRPs, they can be molded to certain shapes without any extra machining, making them ideal for internal aerospace engine components, exhaust systems, and other "hot-zone" structures-even replacing the latest in HRSA metals listed earlier.

4. New Materials Address a New Aerospace Reality

Metallic and composite materials alike continue to be developed and improved to offer ever-increasing performance. Accelerating this evolution of new materials, advancements in machining and cutting technology give manufacturers unprecedented access to materials previously deemed impractical or too difficult to machine. New material adoption is happening exceptionally quickly in aerospace, requiring DFM-minded interaction between material characteristics and component design. The two must be in balance, and one can't really exist outside of the context of the other. Meanwhile, one-piece designs are continuing to reduce the number of components in overall assemblies. A variation of this trend exists in metallic structures, as more components are conditioned in forgings to get to near-net shape, reducing the amount of machining. Elephant skins, roughed-in shapes, and thin floor sections all reduce material costs and the total number of components, but setup and fixturing continue to be challenges. Still, difficulties exist in workholding, surface finish, and CAM tool paths. But designers, machinists, engineers, and machine tool/cutting tool partners are developing new solutions to keep the evolution churning forward.

The mix of materials in aerospace will continue to change in coming years with composites, freshly machinable metals, and new metals increasingly occupying the space of traditional materials. The industry continues to march toward components of lighter weights, increased strengths, and greater heat and corrosion resistance. Component counts will decrease in favor of stronger, near-net shapes, and design will continue its close collaboration with material characteristics. Machine tools builders and cutting tool manufacturers will continue to develop tools to make currently unviable materials machinable, and even practical. And it's all done in the name of reducing the cost of aerospace manufacture, improving fuel economy through efficiency and light weighting, and making air travel a more cost-effective means of transportation.

(Source: *Aerospace materials—past, present, and future*. Aerospace Manufactring and Design Website)

Words and Expressions

state-of-the-art　最先进的，达到最高水准的
jet　*n.* 喷气式飞机，喷射流，喷嘴
fuselage　*n.* ［航］机身（飞机）
airframe　*n.* 商用喷气机机身
actuator　*n.* 制动器，传动装置
aerospace industry　航空航天工业
honeycomb material　蜂窝材料
learn-burn engine　稀薄燃烧发动机
low-pressure turbine　低压涡轮叶片
engine component　发动机部件
fuel efficiency　燃料功率，燃油效率
cruising altitude　巡航高度
landing gear　飞机起落架（着陆装置）
near-net shape　近净成形，维精化

24.3　问题与讨论

24.3　Questions and Discussions

(1) What are important considerations for eliminating degradation by fatigue and corrosion and for reducing in-service maintenance costs?

(2) What is the overall goal of alloy development presently?

(3) Which systems can provide sufficient strength for use in airframes?

(4) Which process can provide higher strength levels in aluminum alloys?

(5) What is always the principal consideration for the selection of aluminum alloys?

单元 25　金属在生物医学中的应用
Unit 25　Metals for Biomedical Applications

扫码查看
讲课视频

25.1　教学内容
25.1　Teaching Materials

1. Introduction

In modern history, metals have been used as implants since more than 100 years ago when Lane first introduced metal plate for bone fracture fixation in 1895. Shortly after the introduction of the 18-8 stainless steel in 1920s, which has had far-superior corrosion resistance to anything in that time, it immediately attracted the interest of the clinicians. Thereafter, metal implants experienced vast development and clinical use. Table 1 summarized the type of metals generally used for different implants division. 316L type stainless steel is still the most used alloy in all implants division ranging from cardiovascular to otorhinology.

Table 1　Implants division and type of metals used

Division	Example of implants	Type of metal
Cardiovascular	Stent Artificial valve	316L SS; CoCrMo; Ti Ti6Al4V
Orthopaedic	Bone fixation (plate, screw, pin) Artificial joints	316L SS; Ti; Ti6Al4V CoCrMo; Ti6Al4V; Ti6Al7Nb
Dentistry	Orthodontic wire Filling	316L SS; CoCrMo; TiNi; TiMo AgSn(Cu) amalgam, Au
Craniofacial	Plate and screw	316L SS; CoCrMo; Ti; Ti6Al4V
Otorhinology	Artificial eardrum	316L SS

Generally, all metal implants are non-magnetic and high in density. These are important for the implants to be compatible with magnetic resonance imaging (MRI) techniques and to be visible under X-ray imaging. Most of artificial implants are subjected to loads, either static or repetitive, and this condition requires an excellent combination of strength and ductility. This is the superior characteristic of metals over polymers and ceramics. Specific requirements of metals depend on the specific implant applications. Stents and stent grafts are implanted to open stenotic blood vessels; therefore, it requires plasticity for expansion and rigidity to maintain dilatation. For orthopaedic implants, metals are required to have excellent toughness, elasticity, rigidity, strength and resistance to fracture. For total joint replacement, metals are needed to be wear resistance; therefore debris formation from friction can be avoided. Dental restoration requires strong and rigid metals and even the shape memory effect for better results.

2. Common Metals Used for Biomedical Devices

Up to now, the three most used metals for implants are stainless steel, CoCr alloys and Ti alloys. The first stainless steel used for implants contains ~18 wt% Cr and ~8 wt% Ni makes it stronger than the steel and more resistant to corrosion. Further addition of molybdenum (Mo) has improved its corrosion resistance, known as type 316 stainless steel. Afterwards, the carbon (C) content has been reduced from 0.08 to 0.03 wt% which improved its corrosion resistance to chloride solution, and named as 316L.

Titanium is featured by its light weight. Its density is only 4.5g/cm^3 compared to 7.9g/cm^3 for 316 stainless steel and 8.3g/cm^3 for cast CoCrMo alloys. Ti and its alloys, i.e. Ti_6Al_4V are known for their excellent tensile strength and pitting corrosion resistance. Titanium alloyed with Ni, i.e. Nitinol, forms alloys having shape memory effect which makes them suitable in various applications such as dental restoration wiring.

CoCr alloys have been utilised for many decades in making artificial joints. They are generally known for their excellent wear resistance. Especially the wrought CoNiCrMo alloy has been used for making heavily loaded joints such as ankle implants (Figure 1).

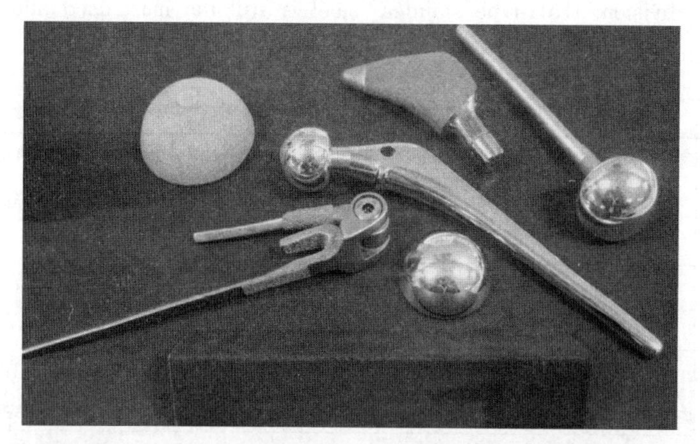

Figure 1　A set of ankle implants (Courtesy of MediTeg, UTM)

3. Biocompatibility of Metals

The understanding of biocompatibility has been focused for long-term implantable devices. With recent development in biotechnology, some level of biological activity is needed in particular research area, such as tissue engineering, drug and gene delivery systems, where direct interactions between biomaterials and tissue components are very essential.

One of the recent definition of biocompatibility is "the ability of a biomaterial to perform its desired function with respect to a medical therapy, without eliciting any undesirable local or systemic effects in the recipient or beneficiary of that therapy, but generating the most appropriate beneficial cellular or tissue response in that specific situation, and optimising the clinically relevant performance of that therapy". In metals, biocompatibility involves the acceptance of an artificial implant by the surrounding tissues and by the body as a whole. The metallic implants do not irritate the surrounding structures, do not incite an excessive inflammatory response, do not stimulate allergic and immunologic reactions, and do not cause cancer. Since metals can corrode in an in vivo environment. Therefore, corrosion resistance of a metallic implant is an important aspect of its biocompatibility.

4. Surface Treatment and Coating

Surface treatment or surface modification is considered as one major concern on recent developments in metallic biomaterials. The treatment includes surface morphological modification and chemical modification. Surface morphology such as roughness, texture and porosity are important characteristics of implant since it influences the ability of cells to adhere to solid substrate. For the case of chemical modification, the objective of the modification is to provide specific biological response on the metallic surface and increase the stability of bio-molecules.

Ti_6Al_4V offers excellent corrosion resistance and ability to be deformed superplastically that make it preferable to substitute complex shape hard tissue. However, Ti_6Al_4V alone does not fully satisfy biocompatibility requirements as implant product. Therefore ceramic bio-apatite such as hydroxyapatite (HAP) or carbonated apatite (CAP), normally are coated on this alloy. The improved biocompatibility is due to the chemical and biological similarity of bio-apatite to hard tissues. Numerous coating methods have been employed to improve bio-compatibility of metal implant including plasma spray and sol-gel. Among the processes, plasma spray has been the most popular method for the coating process of bio-apatite on Ti substrate.

5. Sterilization and Cleaning

In order to avoid bacteria contamination which could be transferred to patients, sterilization and cleaning are important requirements on metal implant. Descaling is a method to clean metal implant surface which can be done mechanically, chemically or by combination of both of the methods. Mechanically it can be done with sand blasting process and chemical cleaning can be done by pickling using strong acid such as NaOH and H_2SO_4. On the other hand sterilization can be done by several processes such as autoclaving, glow discharge Ar plasma treatment and irradiation.

(Source: Reza Fazel-Rezai. *Biomedical Engineering—From Theory to Applications*. Rijeka: Janeza Trdine, 2011)

Words and Expressions

implant *v.* 植入，移植；*n.* 移植物
cardiovascular *a.* [医] 心血管的
otorhinology *n.* 耳鼻科学
stent *n.* 支架
graft *n.* 嫁接，移植
stenotic *a.* 狭窄的
vessel *n.* 血管，船舶，容器
dilatation *n.* 扩张，详述，膨胀度
orthopaedic *a.* 矫形的，整形外科的
joint *n.* 关节，接合处，接缝
debris *n.* 残骸，碎片
restoration *n.* 整修，修复
wrought *a.* 锻造的，加工的，精细的
ankle *n.* 脚踝，踝关节
biocompatibility *n.* 生物相容性
biotechnology *n.* 生物技术

gene *n.* 基因
inflammatory *a.* 炎症性的，煽动性的
allergic *a.* 对……过敏的，过敏性的
immunologic *a.* 免疫学的
cancer *n.* 癌症，痼疾
eliciting *v.* 诱发，刺激，引起
hydroxyapatite *n.* 羟基磷灰石
descaling *n.* 除去锈垢；*v.* 除锈
sterilization *n.* ［医］［食］杀菌
pickling *n.* 酸洗，浸酸
autoclaving *n.* 高压灭菌法；*v.* 用高压灭菌器消毒
bone fracture ［外科］骨折
magnetic resonance imaging 磁共振成像
tissue engineering 组织工程
in vivo （拉）［生物］在活的有机体内
chemical modification 化学修饰，化学改性
sand blasting 喷沙，喷砂清理
plasma treatment 等离子体处理
glow discharge 辉光放电

Phrases

be implanted to	被植入，被移植到
be compatible with	一致；适合；与……相配
be subjected to	受支配；从属于；有……倾向的
adhere to	坚持；黏附；拥护，追随
level of	……的水平；……的等级
the ability of	……的能力

Notes

(1) Specific requirements of metals depend on the specific implant applications. Stents and stent grafts are implanted to open stenotic blood vessels; therefore, it requires plasticity for expansion and rigidity to maintain dilatation.

implant 做名词，尤指医学上的“移植物，植入物”；graft 在生物领域上等同于 implant；be implanted to 译为“被植入，被移植到”。

参考译文：金属的特殊要求取决于具体的植入应用。支架和支架移植物植入用于打开狭窄的血管；因此，它需要一定塑性来进行膨胀，需要一定刚度来保持膨胀度。

(2) With recent development in biotechnology, some level of biological activity is needed in particular research area, such as tissue engineering, drug and gene delivery systems, where direct interactions between biomaterials and tissue components are very essential.

some level of 译为“一定水平的”；where 引导定语从句，修饰 tissue engineering，drug and gene delivery systems。

参考译文：随着生物技术的发展，在特别的研究领域需要一定水平的生物活性，如组织工程、药物和基因递送系统等，其中生物材料与组织成分之间的直接相互作用是非常必要的。

(3) One of the recent definition of biocompatibility is “the ability of a biomaterial to perform its desired function with respect to a medical therapy, without eliciting any undesirable local or systemic effects in the recipient or beneficiary of that therapy, but generating the most appropriate beneficial cellular or tissue response in that specific situation, and optimising the clinically relevant

performance of that therapy".

the ability of 译为"……的能力"；with respect to 译为"关于，在……方面"；此处 but 非转折，是"而是"之意。

参考译文：最近对生物相容性的定义之一是"生物材料在药物治疗方面实现其预期功能的能力，而不会对该治疗的接受者或受益人产生任何不良的局部或全身影响，而是在特定情况下产生最适当的有益细胞或组织反应，并优化该治疗的临床相关性能"。

(4) Surface morphology such as roughness, texture and porosity are important characteristics of implant since it influences the ability of cells to adhere to solid substrate.

"since" 引导原因状语从句；adhere to 译为"黏附，黏到"。

参考译文：表面形态如粗糙度、纹理和孔隙率是植入物的重要特征，因为它影响细胞对固体基质的黏附能力。

25.2　阅读材料

25.2　Reading Materials

New Developments of Ti-Based Alloys for Biomedical Applications

With the development of economy and technology, the number of aged people demanding failed tissue replacement is rapidly increasing. Elderly people have a higher risk of hard tissue failure. It is estimated that 70% ~ 80% of biomedical implants are made of metallic materials. Metallic implants are remarkably important for the reconstruction of failed hard tissue and the market growth rate remains at around 20% and 25%. As human life span grows, the need of biomaterials will definitely continue to increase. This can stimulate the market and research process in a large scale. However, from the application viewpoint, there is still a huge gap between the supply and demand, especially in economically underdeveloped areas where medical technology is very limited. From the research viewpoint, biomaterial is an increasingly important topic, calling for a good mastery of knowledge in materials, biology, physics, chemistry, etc. One feature of biomaterial research is that it has a clear goal and possible applications. In terms of implantation materials, comprehensive properties of low elastic modulus, high strength, excellent wear and corrosion resistance, and good biocompatibility are those characteristics researchers have always been pursuing.

Metals and their alloys are widely used as biomedical materials. On one hand, metallic biomaterials cannot be replaced by ceramics or polymers at present. Because mechanical strength and toughness are the most important safety requirements for a biomaterial under load-bearing conditions, metallic biomaterials like stainless steels, Co-Cr alloys, commercially pure titanium (CP Ti) and its alloys are extensively employed for their excellent mechanical properties. On the other hand, metallic materials sometimes show toxicity and are fractured because of their corrosion and mechanical damages (Figure 1). Therefore, development of new alloys is continuously trialed. Purposes of the development are as follows:

(1) To remove toxic elements;

(2) To decrease the elastic modulus to avoid stress shield effect in bone fixation;

(3) To improve tissue and blood compatibility;

(4) To miniaturize medical devices.

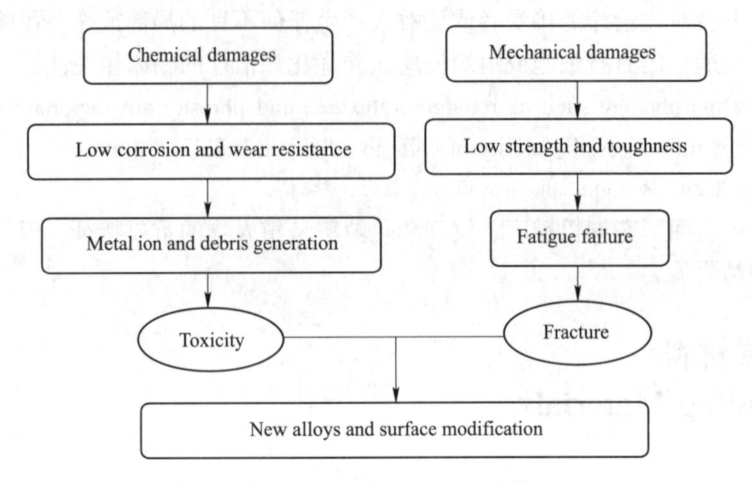

Figure 1　Degradation of metallic materials

The development must be performed on the basis of metallurgy and the resultant alloys must have a good balance between mechanical properties and corrosion resistance. Among metallic materials, titanium and its alloys are considered as the most suitable materials for biomedical applications for their superior comprehensive properties, and they satisfy the requirements of implantation materials better than other competing materials, such as stainless steels, Cr-Co alloys, CP niobium and tantalum. The development of titanium and its alloys used as implant material perfectly reflect the research goal of biomaterials. Firstly, CP Ti was proposed as an alternative for the 316L stainless steel and Co-Cr alloys owing to better biocompatibility and corrosion resistance, since stainless steels and Co-Cr alloys usually contain some harmful elements, such as Ni, Co and Cr. Despite this fact, the mechanical properties of CP Ti cannot satisfy the requirements of biomaterials in some cases when high strength is necessary, such as hard tissue replacement or under intensive wear use. To overcome such restrictions, CP Ti was substituted by $\alpha+\beta$-type Ti-based alloys, particularly Ti-6Al-4V alloy. However, Ti-6Al-4V alloy is composed of cytotoxic elements like Al and V, which may cause severe problems once released inside human body. To overcome the potential V toxicity, V was replaced by Nb and Fe, leading to two new V-free $\alpha+\beta$-type Ti-based alloys, i. e., Ti-6Al-7Nb and Ti-5Al-2. 5Fe. Both alloys show good mechanical and metallurgical behavior comparable to those of Ti-6Al-4V. Nevertheless, several studies have recently shown that the elastic modulus of α-type and some $\alpha+\beta$-type Ti-based alloys is much higher than that of human bone, which can cause stress shielding effect. Therefore, low modulus β-type Ti-based alloys have been extensively developed to alleviate the stress shielding effect. Especially, Ti-Nb-Ta-Zr alloys have much lower elastic modulus of about 48 ~ 55GPa, about half of conventional Ti-6Al-4V alloy. The lowest elastic modulus reported so far in bulk Ti-based alloys developed for biomedical applications is 40GPa for Ti-35Nb-4Sn alloy. However, it is still greater than that of cortical bone (10~30GPa), especially higher than that of cancellous bone with a modulus of about 0.01~2GPa. At present, it is very difficult to lower the elastic modulus of bulk Ti-based alloys below 40GPa.

The main reason why good fixation of implantation materials to the bone tissue remains a problem is the elastic modulus mismatch between biomaterials and the surrounding bones. However, the implanted materials must be strong and durable enough to withstand the physiological loads exerted on it and expected to serve for much longer period or until lifetime without failure or revision surgery. A suitable balance between strength and stiffness to best match that of bone is highly essential.

In order to further reduce elastic modulus of Ti-based alloys, porous materials have been introduced. As we all know, elastic modulus is a property that does not vary easily. The reason why porous materials work is that the amount of materials supporting the same cross section area for porous materials is much less than bulk materials. Thus, if the stress is increased deformation is larger and stiffness is smaller. The main idea of porous alloys is to reduce the stiffness. In addition, porous materials can provide better biological fixation by promoting bone tissue ingrowth into the pores of the implants, which enables homogeneous stress transfer between bones and implants. Presently, porous titanium and its alloys have become an important aspect of biomaterials. They are attracting broad interest from biomaterial researchers. Ti-based biomaterials with tailored porosity are important for cell adhesion, viability, differentiation and growth.

The fabrication of porous materials has been actively investigated since 1943. Sosnik first attempted to introduce pores into Al by adding mercury to the melt. However, porous materials being used as biomaterials have been investigated much later. One of the earliest works that mention the concept of applying porous metals to osseointegration was the work of Weber and White in 1972. Subsequently, numerous researches on porous materials began in the 1970s, including porous ceramics polymers and metallic materials, which were demonstrated to be potential candidates for porous implants in animal experiment. Though porous ceramics and polymers have been studied as scaffold materials, they cannot satisfy requirements under load-bearing conditions. Although ceramics portray excellent corrosion resistance, the porous ceramics might fracture due to intrinsic brittleness. Likewise, porous polymeric systems cannot endure mechanical force present in joint replacement surgery. This impels researchers to focus on porous metals due to their superior mechanical strength and good biocompatibility required for load-bearing applications. Therefore, porous Ti-based alloys are fast emerging as the first choice for biomedical applications. Porous Ti-based alloys exhibit a good combination of mechanical strength with low elastic modulus. Besides, porous structure and rough surface provide better biological fixation and biocompatibility compared with other porous materials. Various methods for fabricating porous Ti-based alloys have been studied recently, including investment casting, sintering loose titanium powder or fibre, slurry sintering, rapid prototyping, sintering a mixture of titanium powder and space holder method. Mechanical properties and architecture of porous Ti-based alloys can be adjusted to be suitable for human bone through the approaches mentioned above. Therefore, porous Ti-based alloys can overcome the mechanical weakness of porous ceramics and polymeric materials as well as eliminating problems of biomechanical mismatch of elastic modulus. At the same time, they possess interconnected structure to provide space for maintenance of stable blood supply and ingrowth of new bone tissues. However, high porosity causes a decrease in mechanical strength of porous materials. In order to achieve a porous biomaterial combined with high strength and high porosity, some new porous Ti-based alloys were developed. New porous Ti-based alloys are expected to combine high mechanical strength with good biocompatibility to best meet the demands of biomedical implants.

(Source: Yuhua Li, et al. *New Developments of Ti-Based Alloys for Biomedical Application*. Materials, 2014, 7, 1709-1800)

Words and Expressions

implantation　*n.* 移植，灌输
cancellous　*a.* 网状骨质的，松质骨
biomaterial　*n.* 生物材料
porosity　*n.* 孔隙率
osseointegration　*n.* 骨结合，骨整合
stress shield effect　应力屏蔽效应
bone fixation　骨固定
cortical bone　皮质骨
porous material　多孔材料
scaffold material　支架材料
investment casting　熔模铸造
slurry sintering　泥浆烧结

25.3　问题与讨论
25.3　Questions and Discussions

(1) What is the superior characteristic of metals over polymers and ceramics for artificial implants?

(2) What is the requirements for orthopaedic implants?

(3) What are the most used metals for implants?

(4) What is "biocompatibility"?

(5) The principle of choosing metallic implants.

(6) What is considered as one major concern on recent developments in metallic biomaterials?

单元 26　金属玻璃
Unit 26　Metallic Glass

扫码查看
讲课视频

26.1　教学内容
26.1　Teaching Materials

1. Introduction

In recent decades, aluminum, steel and plastics have been the most commonly used materials. Plastics are very adaptable materials because of their easy processing, but characterized by low strength compared to metals. On the other hand, aluminum and steel lose the battle with plastics in area of their processing in order to produce very intricate shapes. In this sense, metallic glasses compete with both metals and plastics. These materials have good strength and toughness compared to plastics and can be formed in any desired intricate design compared to metals. Also they possess high corrosion and wear resistance. So, we can say that metallic glasses are the materials having the good properties of metals (like steel and aluminum) as well as good adaptability like plastics.

As promising materials for different applications, these metallic glasses for specific applications are a consequence of combinations of those properties in the followings.

(1) Metallic glasses have no long-range of ordering like crystalline materials. It develops more homogeneity inside the material because defects like point defects, dislocations and stacking faults are absent.

(2) These materials possess very high strength in the elastic region. It can be declared as a good yielding strength of the material which is higher than steel.

(3) Because of the good homogeneity of atoms in metallic glasses, very good corrosion resistance is achieved along with good wear resistance.

(4) Ordinary silica glasses are brittle in nature unlike the metallic glasses which are very tough materials.

(5) These materials have good luster and mirror effects but they are opaque.

(6) The metallic glasses are very hard materials and their fracture resistance is much better compared to ceramics.

(7) Because of the metallic atoms, these glasses possess significant magnetic effects. It helps to easily magnetize or demagnetize these materials. Metallic glasses with soft magnetism have very small hysteresis loop.

(8) Because of the amorphous structure of metallic glasses, their electrical resistivity is higher

resulting in less eddy current loss during its application.

The discovery of metallic glass in 1960 motivated scientists to research and manufacture this kind of materials. They were first manufactured in California Institute of Technology, USA. The researchers got the non-crystalline structure in Au-Si alloys. Rapid quenching of those alloys from their liquid state was conducted by the gun technology. They formed a very thin layer of metallic glass over a cold copper substrate. After that, people are continuously discovering various metallic glasses with different compositions of elements. After 2000 A. D., people are making varieties of metallic glasses and its demand is increasing for industrial applications. In the 1990s, the development of different BMGs (Bulk Metallic Glasses) based on late transition metals (LTM) started. A. Inoue et al. successfully developed the Fe-Al-Ga-P-C-B BMGs in 1995. Today the availability and cost-effectiveness are the two major factors in selecting and production of such materials.

2. Structure, Properties and Applications

Structure of material defines its property. BMGs do not exhibit a long-range order structure, as they solidify from liquid without reaching the crystalline ground state. The structure of the bulk metallic liquids was first observed by Bernal and it was described as dense random packing. Structural features of metallic glasses are discussed by Michael et al. where the concept of efficient filling of space is supported. The rationalization of the good glass forming compositions can be possible by the analysis of dense packing. An example of simple binary metallic glass is shown in Figure 1. These structural motifs arise from the strong tendency to form as many bonds as possible between unlike species because of the large negative heat of mixing which is usual in good glass formers. The replacement of Pt solute in Figure 1 by much smaller Be reduce the number of Zr neighbors which can be accommodated around the solute, and the solute concentration in the alloy would be correspondingly much higher. The medium-range order and dense packing in three-dimensional space can be possible by the overlapping of the cluster via various solvent-atom sharing schemes.

Figure 1　Model of a simple binary metallic glass

The adaptability of the metallic glass in the real world applications is spread in various fields, such as striking face plate in golf clubs, frame in tennis rackets, various shapes of optical mirrors, casing in cellular phones, casing in electro-magnetic instruments, connecting part of optical fibers, shot penning balls, electro-magnetic shielding plates, soft magnetic choke coils, soft magnetic high frequency power coils, high torque geared motor parts, high corrosion resistance coating plates, vessels for lead-free soldering, colliori type liquid flow meter, spring, in-printing plate, high frequency type antenna material, biomedical instruments such as endoscope parts etc. Metallic glasses are very strong compared to other conventional materials and that makes it a very good candidate for military applications like armor (Bulletproof vest) piercing bullets, anti-tank projectiles etc. Those metallic glasses, which are stronger than titanium, are also tried for aerospace application. This type of materials can give relatively double the performance compared to that of a titanium product in the space application. The major problem lies in the BMGs are the very quick aging of these materials.

Firstly China and secondly the United States are the major producers of BMGs. Currently, most applications are focused on electric based products like transformer core. Because of the good conductivity properties of BMGs, it dominates in that sector. Other applications like high-temperature applications, aerospace applications and military applications have a long way to go for becoming a better replacement for the recently used materials.

(Source: Swadhin Kumar Patel, et al. *Metallic Glasses: A Revolution in Material Science.* Rijeka: IntechOpen, 2020)

Words and Expressions

adaptability　*n.* 适应性，可变性，适合性
luster　*n.* 光泽；光彩；*v.* 使有光泽，发亮
motif　*n.* 主题，图形，模体
endoscope　*n.* ［临床］内窥镜
demagnetize　*v.* 去磁
casing　*n.* 包装，外壳
antenna　*n.* 天线
projectile　*n.* 炮弹，抛射体
transformer　*n.* ［电］变压器
metallic glass　金属玻璃
hysteresis loop　［力］滞后回线，磁滞回线
long-range order　长程有序
eddy current loss　涡流损耗
optical fiber　光纤
heat of mixing　混合热

Phrases

characterized by　以……为特征
lose the battle with　输给了……
in this sense　从这个意义上来说
arise from　由……引起，起因于
a consequence of　因此，由于……的结果
go for　去找，被认为，主张，拥护，努力获取

Notes

(1) Plastics are very adaptable materials because of their easy processing, but characterized

by low strength compared to metals. On the other hand, aluminum and steel lose the battle with plastics in area of their processing in order to produce very intricate shapes.

characterized by 译为"以……为特征"; lose the battle with 译为"输给了……"; on the other hand 译为"另一方面"。

参考译文：塑料是一种适应性很强的材料，因为它易于加工，但与金属相比强度较低。另一方面，为生产非常复杂的形状，铝和钢在加工方面输给了塑料。

(2) Structure of material defines its property. BMGs do not exhibit a long-range order structure, as they solidify from liquid without reaching the crystalline ground state.

long-range order 译为"长程有序"; define 译为"决定了，定义了"。

参考译文：材料的结构决定了材料的性质。块体金属玻璃没有长程有序结构，因为它们从液体凝固而没有达到结晶态。

(3) These structural motifs arise from the strong tendency to form as many bonds as possible between unlike species because of the large negative heat of mixing which is usual in good glass formers.

arise from 译为"由……引起，起因于"; the tendency to 译为"……的倾向"; heat of mixing 译为"混合热"。

参考译文：这些结构图案的产生是由于在不同物种间有形成尽可能多成键的强烈倾向，因为良好玻璃前驱体（成形剂）中通常存在较大的负混合热。

(4) Metallic glasses are very strong compared to other conventional materials and that makes it a very good candidate for military applications like armor (Bulletproof vest) piercing bullets, anti-tank projectiles etc.

compared to 译为"与……相比"; conventional 译为"传统的，通用的"。

参考译文：金属玻璃与其他常规材料相比非常坚固，这使得它成为装甲（防弹背心）穿甲子弹、反坦克炮弹等军事应用非常好的候选材料。

(5) Other applications like high-temperature applications, aerospace applications and military applications have a long way to go for becoming a better replacement for the recently used materials.

a long way 译为"很长的路"; go for 译为"去找，被认为，努力获取"。

参考译文：其他应用，如高温应用、航空航天应用和军事应用，要成为最近使用的材料的更好替代品，还有很长的路要走。

26.2　阅读材料

26.2　Reading Materials

Processing of Metallic Glasses

1. General Description of Metallic Glasses

According to atomic arrangement, we can categorize the existing and man-made solid materials into two main groups: crystalline and amorphous. When there is a proper ordered arrangement of atoms then we say it is a crystalline material. If there is a random arrangement of atoms, then the material is called amorphous. To get such randomness, the sizes of the atoms are very important. Much difference in the atomic radius of the components leads to more randomness in the atomic arrangement. Glass forming is majorly concerned with the study of crystallization of materials in order to avoid crystallization. When metallic alloys are cooled at a very fast rate, possibilities of getting an ordered arrangement are poor.

The glass transition temperature (T_g) characterizes amorphous/glass nature of materials. This is more easily understood in the case of a polymer. If we cool a polymer from its liquid state, initially it undergoes cooling and it gets a rubbery state and then after crossing the T_g, it becomes brittle. This kind of phenomenon occurs in amorphous metals too. In case of metallic glasses, we can say that T_g is the temperature at which material gets soft from hard upon heating or get hard upon cooling. This definition for polymers and metals looks similar but it is restricted to amorphous and semicrystalline metals only. The best way to explain the process of getting an amorphous metal or metallic glass is by supercooling the metal from its liquid state. In Figure 1, T_f is the freezing temperature. During cooling, the liquid goes beyond the freezing point and is known as supercooled metal which can have an amorphous structure. In this way, we can get a metallic glass. In the absence of supercooling, the liquid has a tendency to crystallize. The difference in Gibbs free energy between liquid and crystalline state is an important factor for the ability of a metal to crystallize or to become amorphous. Therefore, during the manufacturing of the metallic glasses, kinetics of the supercooling has a great impact on the quality of the glasses.

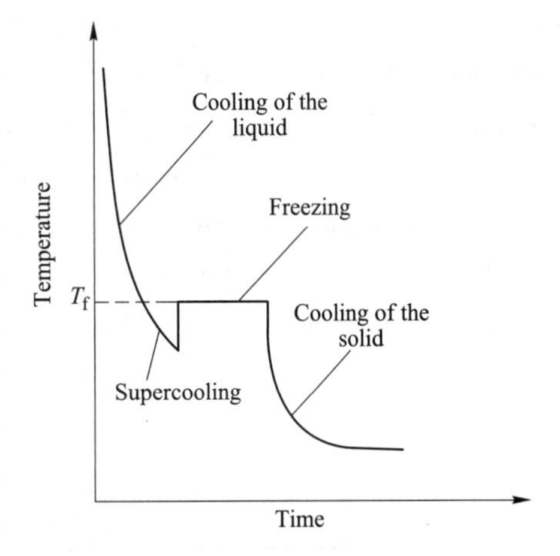

Figure 1 Principle of supercooling

Considering the periodic table, the metallic glasses are mainly divided into two categories: metal-metal and metal-metalloid. As shown in the periodic table, metals can be alkali, alkaline and rare earth metals etc. In metal-metal type, the atomic percentage of individual constituents can be up to 50%. In the case of metal-metalloid type, metalloids like B, Si, Ge can be mixed with metals like Fe, Ni, Co etc. The composition percentage of metalloids in this category is lower than percentage of metals. After the discovery, the compositions of metalloids in the glasses were generally up to 20% but gradually people work on that problem and successfully decreased its percentage beyond 20%. This type of glasses is more often used in commercial applications.

2. Processing of Metallic Glasses

The flow chart for the processing of metallic glasses in order to obtain different products is presented in Figure 2.

(1) Direct Casting

For the net shape fabrication of BMG, two types of casting processes (suction and die casting) have been adopted. Among the two methods the suction casting method develops a product with higher quality and lower porosity than die casting method. BMGs having low melting temperature are beneficial because this process reduces tool cost and wear, lowers energy consumption and shortens cycle time. During the casting processes (both die casting and suction casting) shrinkage has to be taken into consideration. The shrinkage phenomenon is absent in the BMGs formers due to the absence of a first order phase transition during solidification. Cooling process in the casting plays a vital role. Low solidification shrinkage in BMGs develops a gap between the mold and the BMG during the cooling process. The heat transfer through the gap is different for the presence of atmosphere and vacuum, can become the rate limiting factor and leads to affect the cooling rate. In the process of direct casting of BMG formers, fast cooling and forming have to be done simultaneously due to the crystallization mechanism and the crystallization kinetics. Direct casting needs care during the mold filling and at the same time to avoid crystallization during solidification. Figure 3 represents some application of BMGs fabricated by direct casting. The advantages and disadvantages of the direct casting BMG process are given in Table 1.

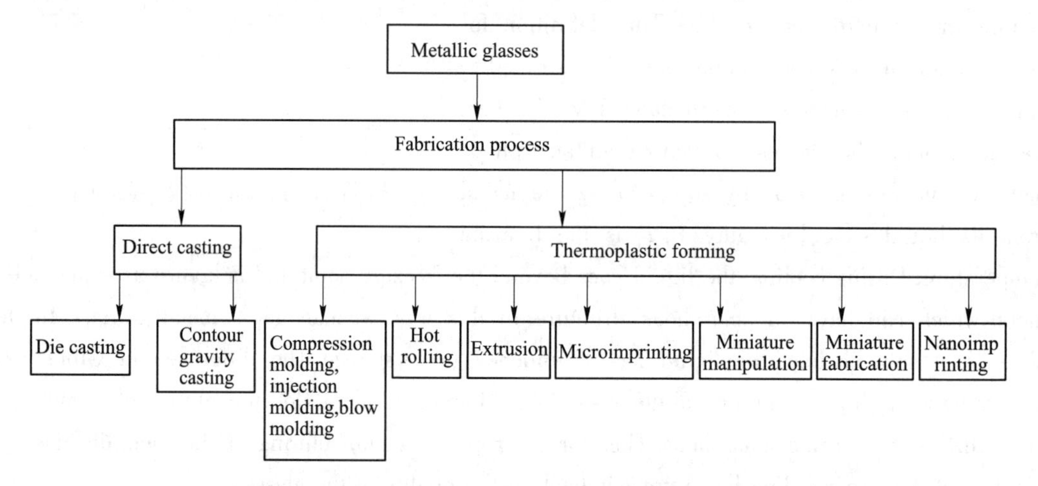

Figure 2　Flow chart of the processing of metallic glasses

Table 1　Advantages and disadvantages of the direct casting process

Advantages	Disadvantages
• Low melting temperature • Low shrinkage • One step process • Homogeneous microstructure • Mechanical properties are already matured in the as-cast state	• Cooling and forming are coupled • Processing environment can influence crystallization kinetics • High viscosity • Internal stresses • BMGs contaminate during processing

(2) Thermoplastic Forming

Thermoplastic forming (TPF) is the alternative process to direct casting for developing BMGs. This process has different nomenclature such as hot forming, hot pressing, superplastic forming, viscous flow working and viscous flow forming. The preferable condition for TPF is the drastic softening of the BMG former upon heating above T_g and its thermal stability. The measure of the ability of BMG

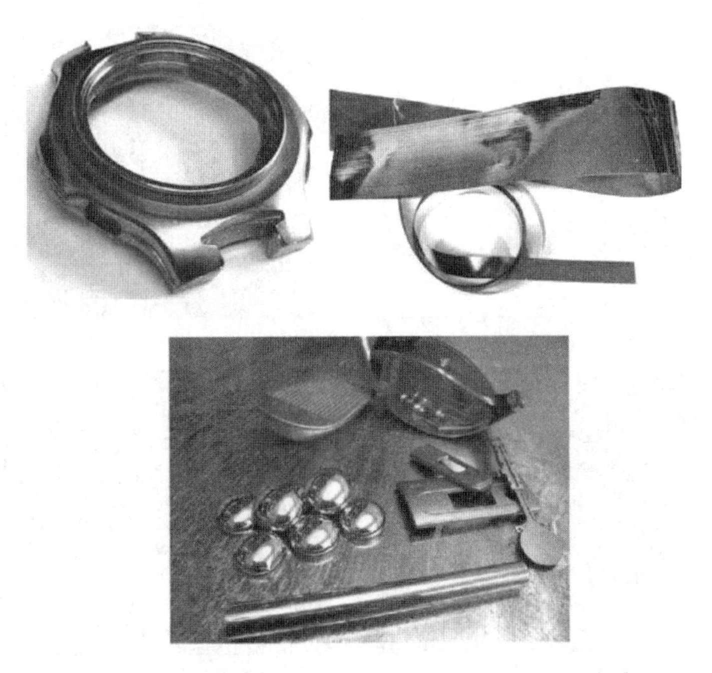

Figure 3 BMG articles fabricated by direct casting method

formers to adopt an amorphous structure by heating above its glass transition temperature is known as thermal stability and it can be quantified by the width of the supercooled liquid region(SCLR).

1) TPF Based Compression and Injection Molding

Compression molding was an adopted form of plastic processing. In this process, the feedstock material is placed in a mold and is given temperature into SCLR and pressure must exceed the flow stress of the BMG to achieve the required strain prior to crystallization setting. Fast cooling is not required for the forming. Figure 4 represents the schematic diagram of the process equipped with examples. Under low applied pressure in the compression molding Pd-Ni-Cu-P alloy can be formed as a gear-shaped structure. It can be noticed that a dense compact part with mechanical properties close to those of bulk material and an outstanding surface finish can be obtained. The needed molding pressure not only depends on the formability of the BMG at the processing temperature but also on the shape of the final product. Injection molding is also a TPF based molding process.

2) Miniature Fabrication

The development in technologies like micro-electromechanical systems (MEMS), electronics devices, and medical devices have created a rising demand for miniature products and parts. The miniature formation is done by different processes like German method LIGA (lithography, electroplating and molding), UV-LIGA etc. Both processes can be used as the surface patterning techniques for creating high aspect ratio structures. Due to the homogeneous and isotropic structure of BMGs in atomic scale and their superior properties over conventional materials used for miniature applications and the capability to produce stress-free parts, these methods attracted a lot of attention.

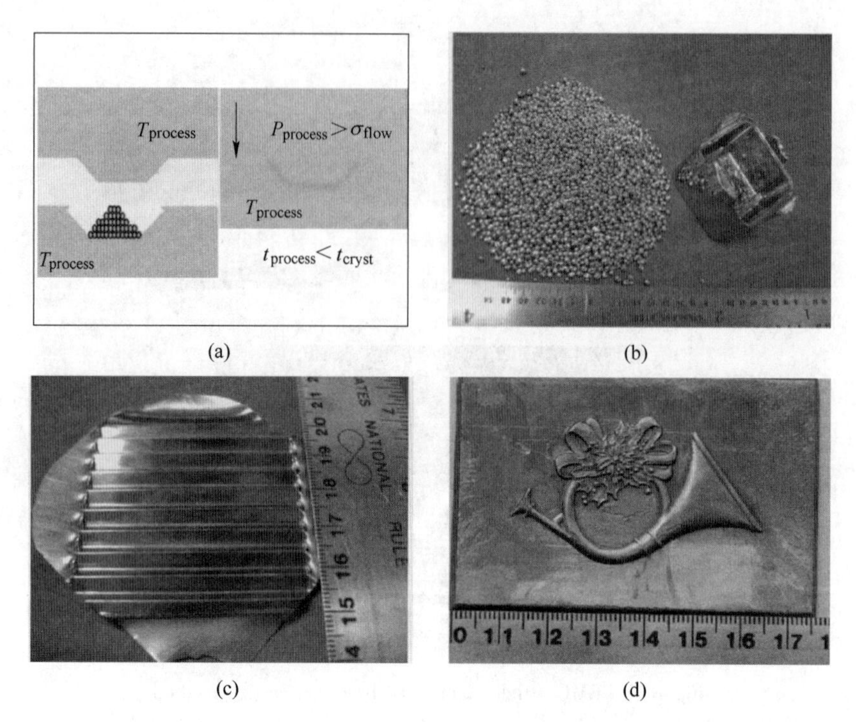

Figure 4 TPF based compression molding with BMG

(a) Schematic diagram of the compression molding with BMGs;

(b) Pellets used as feedstock material to compression mold $Pt_{57.5}Cu_{14.7}Ni_{5.3}P_{22.5}$;

(c) $Zr_{44}Ti_{11}Cu_{10}Ni_{10}Be_{25}$ formed from a flat plate into a corrugated structure;

(d) $Zr_{44}Ti_{11}Cu_{10}Ni_{10}Be_{25}$ formed from a flat plate to create an embossing mold

3) Nano Forming

BMGs basically formed by top-down nanofabrication. The combination of different properties like high strength at room temperature, the ability to imprint a nanometer-sized parallel print process, the non-linear softening of BMGs when reaching their glass transition and the ability to repeatedly write and erase facility on the BMG surface recommends a wide range of application. Nanoimprinting on BMG permits to directly write, such as with atomic force microscopy tip, as in a scanning probe lithography process. The capability of BMGs for direct nanoimprinting can be applied with a combination of surface smoothening method and used as a rewritable high-density data storage. Several materials have been developed for the mold formation and imprint for nanoforming such as silicon, quartz, and alumina.

4) Rolling

Rolling of metallic glasses can be categorized into two processes; one is based on liquid processing and the other is on thermoplastic forming. The example of the former is the melt spinning. In the melt spinning, the liquid sample is quenched by injecting on a single, fast-spinning copper roll. Another process is cold rolling where the BMGs are rolled at the room temperature.

5) Extrusion

The thermoplastic formability of BMG formers can also be used for extrusion. The major advantage of extrusion is that it produces the highest aspect ratio shapes, 100 +, of uniform cross-section. During extrusion after the outlet of the effective die length, the swelling of the material is a common phenomenon (Table 2).

Table 2　Advantage and disadvantages of TPF based BMG processing

Advantages	Disadvantages
• Forming and fast cooling decoupled • Highest dimensional accuracy • Insensitive to heterogeneous influence • Novel and unique process • Low capital investment	• Two or more step process • Novel and unique process

3. Conclusions

BMGs are now available in many different chemical compositions. Although various routes of their processing were adopted, still there is much scope for inventions targeting the development of new processing techniques and new types of glasses. TPF-based processing technique is accepted as a more suitable manufacturing process for obtaining glassy products. This process is competing with the processing of plastics. Very intricate shapes of different geometry, almost unachievable for metals, can be easily achieved with BMGs using TPF-based processing.

(Source: E A Davis, I M Ward. *An Introduction to Matal Matrix Composites*)

Words and Expressions

suction　*n.* 吸模
lithography　*n.* 刻蚀，平版印刷，光刻
electroplating　*n.* 电镀，电镀术
micro-electromechanical　*n.* 微机电系统
isotropic　*a.* ［物］各向同性的，等方性的
viscous　*a.* 黏性的，黏的
glass transition temperature　玻璃化转变温度
thermoplastic forming　热塑性成型
die casting　压铸
superplastic forming　超塑性成型
melt spinning　熔融纺丝

26.3　问题与讨论
26.3　Questions and Discussions

(1) Why metallic glasses can compete with both metal and plastics?
(2) Combination properties of metallic glasses.
(3) What are the two major factors in selecting and producting of such materials?
(4) What is the structure of metallic glasses?
(5) Introduce some real-world applications of metallic glasses.

单元 27 金属基复合材料
Unit 27 Metal Matrix Composites

扫码查看
讲课视频

27.1 教学内容
27.1 Teaching Materials

1. Types of MMC and General Microstructural Features

The term metal matrix composite (MMC) encompasses a wide range of scales and microstructures. Common to them all is a contiguous metallic matrix. The reinforcing constituent is normally a ceramic, although occasionally a refractory metal is preferred. The composite microstructures may be subdivided, as depicted in Figure 1, according to whether the reinforcement is in the form of continuous fibres, short fibres or particles.

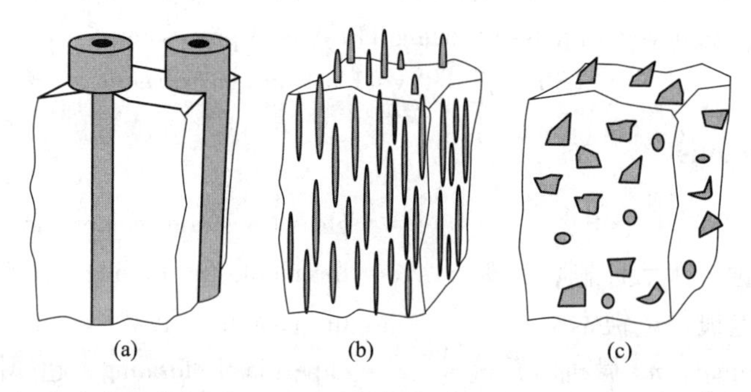

Figure 1 Schematic depiction of the three types of MMC, classified according to the type of reinforcement
(a) Monofilaments; (b) Whiskers/Staple fibres; (c) Particulate

2. Historical Background

Examples of metal matrix composites stretch back to the ancient civilisations. Copper awls from Cayonu (Turkey) date back to about 7000 B. C. and were made by a repeated lamination and hammering process, which gave rise to high levels of elongated non-metallic inclusions. Among the first composite materials to attract scientific as well as practical attention were the dispersion hardened metal systems. These developed from work in 1924 by Schmidt on consolidated mixtures of aluminium/alumina powders and led to extensive research in the 1950s and 1960s.

For both dispersion hardening and precipitation hardening, the basis of the strengthening mechanism is to impede dislocation motion with small particles. This is achieved by the incorporation of either fine oxide particles or non-shearable precipitates within a metallic matrix. Of prime importance in this context is the minimisation of the spacing between the inclusions. Since it is generally possible to achieve finer distributions in precipitation hardened systems, these normally exhibit higher strengths at room temperature. However, dispersion strengthened systems show advantages at elevated temperature, because of the high thermal stability of the oxide particles.

More recent developments have brought the concept of metal matrix composites closer to engineering practice. An interesting example is provided by the so-called "dual phase" steels, which evolved in the 1970s. These are produced by annealing fairly low carbon steels in the $\alpha+\gamma$ phase field and then quenching so as to convert the γ phase to martensite. The result is a product very close to what is now referred to as a particulate MMC, with about 20% of very hard, relatively coarse martensite particles distributed in a soft ferrite matrix. This is a strong, tough and formable material, now used extensively in important applications such as car bodywork.

Interest in fibrous metal matrix composites mushroomed in the 1960s, with effort directed mainly at aluminium and copper matrix systems reinforced with tungsten and boron fibres. In such composites the primary role of the matrix is to transmit and distribute the applied load to the fibres. Research on continuously reinforced composites waned during the 1970s, largely for reasons of high cost and production limitations. Discontinuously reinforced composites fall somewhere between the dispersion strengthened and fibre strengthened extremes, in that both matrix and reinforcement bear substantial proportions of the load. They have been rapidly developed during the 1980s, with attention focused on Al-based composites reinforced with SiC particles and Al_2O_3 particles and short fibres. The combination of good transverse properties, low cost, high workability and significant increases in performance over unreinforced alloys has made them the most commercially attractive system for many applications.

3. Interactions Between Constituents and the Concept of Load Transfer

One definition of the word "composite" is simply: "something combining the typical or essential characteristics of individuals making up a group". Central to the philosophy behind the use of any composite material is the extent to which the qualities of two distinct constituents can be combined, without seriously accentuating their shortcomings. In the context of MMCs, the objective might be to combine the excellent ductility and formability of the matrix with the stiffness and load-bearing capacity of the reinforcement, or perhaps to unite the high thermal conductivity of the matrix with the very low thermal expansion of the reinforcement.

In attempting to identify attractive matrix/reinforcement combinations, it is often illuminating to derive a "merit index" for the performance required, in the form of a specified combination of properties. Appropriate models can then be used to place upper and lower bounds on the composite properties involved in the merit index, for a given volume fraction of reinforcement. Central to an understanding of the mechanical behaviour of a composite is the concept of load sharing between the matrix and the reinforcing phase. The stress can vary sharply from point to point, but the proportion of the external load borne by each of the individual constituents can be gauged by volume-averaging the load within them. The reinforcement may be regarded as acting efficiently if it carries a relatively high proportion of the externally applied load. This can result in higher strength, as well as greater stiffness, because the reinforcement is usually stronger, as well as stiffer, than the matrix.

(Source: Clyne T W, et al. *An Introduction to Metal Matrix Composites*. Cambridge: Cambridge University Press, 1993)

Words and Expressions

feature　*n.* 特征，方面，特写
contiguous　*a.* 连续的，邻近的，接触的
constituent　*n.* 组成，成分，构成要素
awl　*n.* 锥子，尖钻
bodywork　*n.* 车体，车身制造
wane　*v.*（月亮）缺，减少；*n.* 减弱
accentuate　*vt.* 强调，重读
mushroomed　*a.* 辐射环式的；*v.* 使迅速成长
boron　*n.*［化］硼
incorporation　*n.* 掺入，掺杂
gauge　*n.* 测量仪器，宽厚度；*v.* 估计，测量
metal matrix composite　金属基复合材料
dispersion hardering　弥散硬化
thermal expansion　［热］热膨胀
load transfer　荷载传递，负荷转移
load sharing　荷载分担，负载分配

Phrases

in the form of　以……的形式
stretch back to　回溯到，起源于
a high proportion of　一大部分，高比例的
date back to　追溯到；从……开始有
give rise to　导致，得以兴起

Notes

(1) This is achieved by the incorporation of either fine oxide particles or non-shearable precipitates within a metallic matrix. Of prime importance in this context is the minimisation of the spacing between the inclusions.

the incorporation of 译为“掺入，掺杂”；fine 译为“精细的，细小的”；of prime importance 译为“最重要的”。

参考译文：这是通过在金属基体中加入细小的氧化物颗粒或不可剪切的沉淀物来实现的。在这种情况下，最重要的是尽量减小夹杂物之间的间距。

(2) The result is a product very close to what is now referred to as a particulate MMC, with about 20% of very hard, relatively coarse martensite particles distributed in a soft ferrite matrix.

close to 译为“接近于，在附近”；refer to 译为“所提及的，指的是”。

参考译文：结果是一种非常接近于现在所称的粒状 MMC 材料，大约有 20%的非常硬的、相对粗大的马氏体颗粒分布在软的铁素体基体中。

(3) The combination of good transverse properties, low cost, high workability and significant increases in performance over unreinforced alloys has made them the most commercially attractive system for many applications.

the combination of 译为“结合，综合”；attractive for 译为“对……有吸引力”。

参考译文：良好的横向性能、低成本、高可加工性以及与未增强合金相比性能的显著提高使其成为许多应用中最具商业吸引力。

(4) Central to the philosophy behind the use of any composite material is the extent to which

the qualities of two distinct constituents can be combined, without seriously accentuating their shortcomings.

which 引导宾语从句；accentuating 译为“加重，强调”。

参考译文：使用任何复合材料背后的理念的核心是在不加重其缺陷情况下可以将两种不同成分的材料进行组合的程度。

(5) The reinforcement may be regarded as acting efficiently if it carries a relatively high proportion of the externally applied load. This can result in higher strength, as well as greater stiffness, because the reinforcement is usually stronger, as well as stiffer, than the matrix.

regards as 译为“把……认作”；a high proportion of 译为“一大部分，高比例的”；reinforcement 译为“强化，增强剂，强化剂”；if 引导条件状语；because 引导原因状语从句。

参考译文：如果钢筋承受大部分的外加载荷，则可认为其有效发挥作用。这会导致更高的强度和更大的刚度，因为钢筋通常比基体更坚固，也更坚硬。

27.2　阅读材料
27.2　Reading Materials

Aluminium Metal Matrix Composites—A Review

1. Introduction

MMC (Metal matrix composites) are metals reinforced with other metal, ceramic or organic compounds. They are made by dispersing the reinforcements in the metal matrix. Reinforcements are usually done to improve the properties of the base metal like strength, stiffness, conductivity, etc. Aluminium and its alloys have attracted most attention as base metal in metal matrix composites. Aluminium MMCs are widely used in aircraft, aerospace, automobiles and various other fields. The reinforcements should be stable in the given working temperature and non-reactive too. The most commonly used reinforcements are Silicon Carbide (SiC) and Aluminium Oxide (Al_2O_3). SiC reinforcement increases the tensile strength, hardness, density and wear resistance of Al and its alloys. The particle distribution plays a very vital role in the properties of the Al MMC and is improved by intensive shearing. Al_2O_3 reinforcement has good compressive strength and wear resistance. Boron Carbide is one of hardest known elements. It has high elastic modulus and fracture toughness. The addition of Boron Carbide (B_4C) in Al matrix increases the hardness, but does not improve the wear resistance significantly. Fibers are the important class of reinforcements, as they satisfy the desired conditions and transfer strength to the matrix constituent influencing and enhancing their properties as desired. Zircon is usually used as a hybrid reinforcement. It increases the wear resistance significantly. In the last decade, the use of fly ash reinforcements has been increased due to their low cost and availability as waste by-product in thermal power plants. It increases the electromagnetic shielding effect of the Al MMC.

2. Silicon Carbide Reinforced AMC

Tamer Ozbenet investigated the mechanical and machinability properties of SiC particle reinforced

Al-MMC. With the increase in reinforcement ratio, tensile strength, hardness and density of Al MMC material increased, but impact toughness decreased. Sedat Ozdenet investigated the impact behaviour of Al and SiC particle reinforced with AMC under different temperature conditions. The impact behaviour of composites was affected by clustering of particles, particle cracking and weak matrix-reinforcement bonding. The effects of the test temperature on the impact behaviour of all materials were not very significant. Srivatsan et al. conducted a study of the high cycle fatigue and investigated the fracture behaviour of 7034/SiC/15p-UA and 7034/SiC/15p-PA metal matrix composites. The modulus, strength and the ductility of the two composite microstructures decreased with an increase in temperature. The degradation in cyclic fatigue life was more pronounced for the under-aged microstructure than the peak-aged microstructure. Maik Thunemann studied the properties of AMMC's based on preceramic-polymer-bonded SiC performs. Polymethylsiloxane (PMS) was used as a binder. A polymer content of 1.25% by weight conferred sufficient stability to the preforms to enable composite processing. It is thus shown that the PMS derived binder confers the desired strength to the SiC preforms without impairing the mechanical properties of the resulting Al/SiC composites.

3. Aluminium Oxide Reinforced AMC

Park investigated the effect of Al_2O_3 in Aluminium for volume fractions varying from 5% ~ 30% and found that the increase in volume fraction of Al_2O_3 decreased the fracture toughness of the MMC. This is due to decrease in inter-particle spacing between nucleated micro voids. Tjong compared the properties of two aluminium metal matrix Composites, Al-B_2O_3-TiO_2 system and Al-B-TiO_2 system. It was found that the reactive hot pressing of the composites resulted in the formation of ceramic Al_2O_3 and TiB_2 particulates as well as coarse intermetallic Al_3Ti blocks. Al-B-TiO_2 and higher Al_3Ti content and showed high tensile strength, but low tensile ductility. Al-B_2O_3-TiO_2 had more fatigue strength than Al-B-TiO_2. Abhishek Kumar et al. experimentally investigated the characterization of A359/Al_2O_3 MMC using electromagnetic stir casting method. They found that the hardness and tensile strength of MMC increases and electromagnetic stirring action produces MMC with smaller grain size and good particulate matrix interface bonding.

4. Boron Carbide Reinforced AMC

Bo Yao. investigated the trimodal aluminium metal matrix composites and the factors affecting its strength. The test result shows that the attributes like nano-scale dispersoids of Al_2O_3, crystalline and amorphous AlN and Al_4C_3, high dislocation densities in both NC-Al and CG-Al domains, interfaces between different constituents, and nitrogen concentration and distribution leads to increase in strength. Vogt studied the cryomilled aluminium alloy and boron carbide nano-composite plates made in three methods, (1) hot isostatic pressing (HIP) followed by high strain rate forging (HSRF), (2) HIP followed by two-step quasi-isostatic forging (QIF), and (3) three-step QIF. The test results showed that the HIP/HSRF plate exhibited higher strength with less ductility than the QIF plates, which had similar mechanical properties. Barbara Previtali investigated the effect of application of traditional investment casting process in aluminium metal matrix composites. Aluminium alloy reinforced with SiC and B_4C were compared and the experiments showed the wear resistance of SiC reinforced MMC is higher than that of B_4C reinforced MMC.

5. Fiber Reinforced AMC

Sayman studied the elasto plastic stress analysis of aluminium and stainless steel fiber and found that under 30MPa pressure and at a temperature of 600℃, good bonding between matrix and fiber was observed, moreover increase in the load carrying capacity of the laminated plate was also visualised. Onur Sayman analysed the elastic-plastic thermal stress on steel fiber reinforced Aluminium metal-matrix composite beams and found that the intensity of the residual stress and the equivalent plastic strain are greatest at 0° orientation angle and concluded that the higher the orientation angle the lower the temperature that causes plastic yielding. Ding et al. investigated the low cycle fatigue behaviour of the pure Al reinforced with 20% Al_2O_3 fiber in total strain controlled mode. They found that the predicted fatigue lives coincide with the observed fatigue lives over a wide range of strain amplitudes for a wide range of test temperatures. Ding investigated the behaviour of the unreinforced 6061 aluminium alloy and short fiber reinforced 6061Al alloy MMC. They found that the addition of high strength Al_2O_3 fibres in the 6061 aluminium alloy matrix will not only strengthen the microstructure of the 6061 aluminium alloy, but also channel deformation at the tip of a crack into the matrix regions between the fibres and therefore constrain the plastic deformation in the matrix which leads in reduction of fatigue ductility.

6. Zircon Reinforced AMC

Jenix Rina compared the properties of Al6063 MMC reinforced with Zircon Sand and Alumina. The hardness and the tensile strength of the composites are higher for volume fractions of Zircon sand and Alumina(4+4)%. In this combination, the particle dispersion is uniform and the pores are less where inter-metallic particles are formed. Sanjeev Das comparatively studied the abrasive wear of Al-Cu alloy with alumina and Zircon sand particles and found that wear resistance of the alloy increases significantly after the addition of alumina and zircon particles. However, zircon reinforced composites showed better wear resistance than that of alumina reinforced composite. Scudino et al. investigated the mechanical properties of Al-based metal matrix composites reinforced with Zircon-based glassy particles produced by powder metallurgy. The test results showed that the compressive strength of pure Al increases by 30% with 40% volume of glass reinforcement. While the volume fraction of the glassy phase increasing to 60%, the compressive strength further increases by about 25%.

7. Fly Ash Reinforced AMC

Fly ash particles are potential discontinuous dispersoids used in metal matrix composites due to their low cost and low density reinforcement which are available in large quantities as a waste by product in thermal power plants. The major constituents of fly ash are SiO_2, Al_2O_3, Fe_2O_3, and CaO. Rajan et al. compared the effect of the three different stir casting methods on the properties of fly ash particles reinforced Al-7Si-0. 35Mg alloy. The three stir casting methods are liquid metal stir casting, compocasting, modified compocasting followed by squeeze casting. The compression strength of the composite processed by modified compocasting cum squeeze casting is improved compared to the matrix alloy. However, the tensile strength was found to be reduced. The modified compocasting cum squeeze casting process has resulted in a well dispersed and porosity free fly ash particle dispersed composite. Zuoyong Dou studied the electromagnetic interference shielding effectiveness properties of the 2024 Al alloy fly ash composites. The composite has effective shielding property in the frequency range of 30. 0Hz-1. 5GHz. But the addition of fly ash particulate decreases the tensile strength of the composites.

8. Summary

Several confronts must be surmounted in order to strengthen the engineering usage of AMCs such as processing methodology, influence of reinforcement, effect of reinforcement on the mechanical properties and its corresponding applications. The major conclusions derived from the prior works carried out can be summarised as below.

(1) SiC reinforced Al MMCs have higher wear resistance than Al_2O_3 reinforced MMCs.

(2) It has been found that the increase in volume fraction of Al_2O_3 decreases the fracture toughness of the Al MMC.

(3) The optimum conditions for fabricating Al_2O_3 reinforced Al MMC as pouring temperature −700℃, preheated mould temperature −550℃, the stirring speed −900rev/min, particle addition rate −5g/min, the stirring time −5min and the applied pressure was 6MPa.

(4) The wear resistance of SiC reinforced Al MMC is higher than B_4C reinforced MMC.

(5) Al MMCs reinforced with diamond fiber exhibit high thermal conductivity and a low thermal expansion coefficient.

(6) The wear resistance and compressive strength of Al MMCs increase with the addition of zircon sand reinforcement.

(7) The addition of fly ash reinforcement in Al increases the wear resistance but decreases the corrosion resistance.

(Source: B Vijaya Ramnath, et al. *Aluminium Metal Matrix Composites—A Review*. Review on Advanced Material Science, 2014, 38: 55-60)

Words and Expressions

polymethylsiloxane　*n.* 聚二甲硅氧烷
dispersoid　*n.* 分散体
cryomilled　*n.* 冷冻研磨
compocasting　*n.* 复合铸造法
intensive shearing　高速剪切，剧烈剪切
hot isostatic pressing　热等静压

27.3　问题与讨论
27.3　Questions and Discussions

(1) What is the normal reinforcing constituent of MMC?

(2) Introduce the forms of reinforcement.

(3) What is the strengthening mechanism of both dispersion hardening and precipitation hardening?

(4) What is the primary role of matrix?

(5) What is the central to the philosophy behind the use of any composite material?

(6) What is the central to an understanding the mechanical behavior of a composite?

单元 28　金属的未来
Unit 28　The Future of Metals

扫码查看
讲课视频

28.1　教学内容
28.1　Teaching Materials

On 15 December 2009, the world's most fuel-efficient commercial jetliner—the Boeing 787 Dreamliner—completed its first flight. The airliner is mostly made from carbon fiber-reinforced polymeric composites (50% by weight, up from 12% in the Boeing 777). Traditional metals are substantially replaced by composites with higher strength/weight ratios; aluminum usage has dropped to 20% (versus 50% in the 777). Ever since the 1950s, when "engineering materials" mainly meant metals, the share of metals in engineering materials has been diminishing. What are the reasons behind this trend, and which applications are likely to stay in the domain of metals?

The main property limitation of metals as structural materials is their low specific strength. Most engineering designs call for structural materials that have high strength, fracture toughness, and stiffness while minimizing weight. This is a key reason for replacing metals in aircraft and sporting goods, where weight is a primary concern. Some metals such as aluminum and magnesium are light, but they are too soft for many applications and have low toughness and stiffness. Titanium alloys partly overcome these problems: They are about half as dense as steels, have higher strength, and are very tough. Titanium was first used in airliners in the 1960s in the Boeing 707 and its use has increased to 15% in the Boeing 787.

Metals can be strengthened through controlled creation of internal defects and boundaries that obstruct dislocation motion. But such strategies compromise ductility and toughness. One method for strengthening metals without losing toughness is grain refinement, but when the grain sizes fall below ~1μm, strengthening is usually accompanied by a drop in ductility and toughness. A recent study points the way to overcoming this problem: In a low-alloy steel containing ultrafine elongated ferrite grains strengthened with nanosized carbides, toughness and strength both rose when temperature was lowered from 60℃ to −60℃. Nanotwinned metals are another example of hierarchical nanostructured metals with extraordinary mechanical properties. When a high density of twin boundaries is incorporated into polycrystalline copper grains, with boundary spacing in the nanometer scale, the material becomes stronger than coarse grained copper by a factor of 10; it is also very ductile. The ultrastrong nanotwinned copper has an electrical conductivity comparable to that of high-conductivity copper and a much enhanced resistance against electromigration. It has great potential for applications in microelectronics.

Corrosion is another headache for metals. To protect metals from corrosion, they are commonly coated with a layer of corrosion-resistant material. The Hangzhou Bay Bridge in China is an outstanding example of this technique. The tubes are protected against corrosion in the harsh ocean environment by a coating of novel polymeric composites combined with cathode attachments. Metal corrosion can also be resisted by forming a continuous protective passivation layer on the metal surface. For example, Yamamoto have added 2.5% Al to conventional austenitic stainless steels, resulting in the formation of a protective aluminum oxide layer that can resist further oxidation at elevated temperatures. Another route to enhancing the corrosion resistance of metals is to modify the chemical composition of surface layers. But most such processes require high temperatures that may cause serious deterioration of the metal substrates.

Metals also get soft at elevated temperatures; they can rarely be used above 1000℃ with useful strengths. Superalloys have higher operating temperatures (up to 1150℃ for nickel-based superalloys), enabling their use in high-temperature applications such as jet turbine engines. Superalloys for higher operating temperatures, based on metals with higher melting points such as molybdenum and niobium, are under investigation.

Despite these limitations, metals are still the major workhorse of our society and will remain so in the future, thanks to unique properties that make them irreplaceable.

First, metals have a much higher fracture toughness than other materials (Figure 1); steels are the toughest known materials. Therefore, metals are usually used for key components with the highest requirements for reliability and durability, such as bridge cables, concrete reinforcement in buildings, and vehicle body frames.

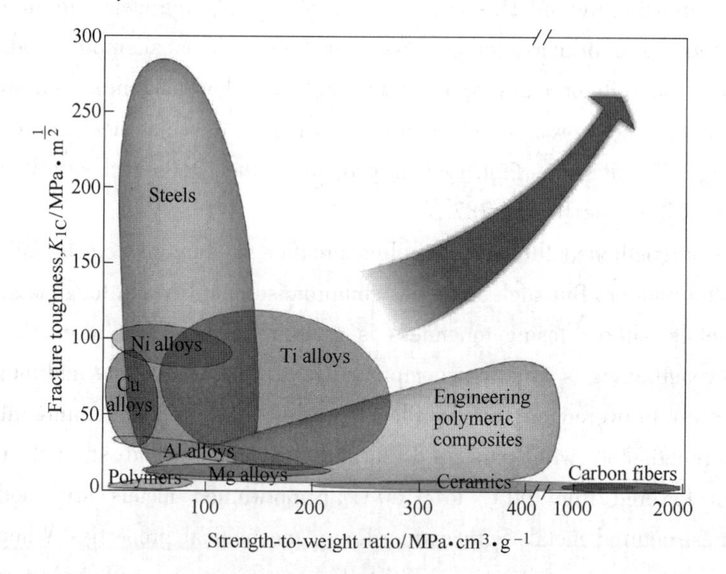

Figure 1　Competing materials

(Steels have the highest toughness, whereas carbon fibers have the highest strengths)

Second, the properties of metals are uniform in all directions, and their strength is the same in tension and compression. Also, the strengths of metals are usually predictable. These features are critically important for predicting fracture in engineering structures. In contrast, it remains very difficult to predict fracture of composites and ceramics: their fracture is often catastrophic and may cause serious economic loss and even loss of life. Hence, many advanced technologies continue to rely on high-performance metals.

Third, most metals are more conductive than ceramics and polymers. Copper and aluminum remain the best materials for overhead electricity transmission lines. Conducting lines and thermal spreaders used in information technology are mostly made from copper and its alloys. Metals also have unique magnetic properties that are not easily reproduced in other materials.

Fourth, metals have the best overall mechanical properties at temperatures up to a few hundred degrees. This window covers most of the operation temperatures in chemical engineering processes, power stations, and various engines. Finally, most metals are recyclable, making metals more competitive for quantity applications.

Modern technologies not only strongly rely on these unique properties of metals, but urgently call for even better metals. Increasing the strength of metals without sacrificing other properties is critical for their competitiveness. Multiscale hierarchical structures provide a possible route to optimizing overall properties. Metals may also be mixed with other materials in a controlled way to form composite structures. Assembling metals with other components in this way may shift their strength/toughness ratios toward the upper right corner in the figure (Figure 1).

(Source: K Lu. *The Future of Metals*. Science, 2010, 328: 319)

Words and Expressions

ultrastrong *a.* 超强的
electromigration *n.* 电迁移
modify *v.* 改性，修饰，使缓和
hierarchical *a.* 分等级的，等级制度的
superalloy *n.* 超耐热，高应力，耐蚀合金
workhorse *n.* 重负荷机器；*a.* 工作重的
reinforcement *n.* 增强，强化
carbon fiber 碳纤维
polymeric composite 聚合物复合材料
specific strength 比强度（强度/重量比）
passivation layer 钝化层，保护层

Phrases

incorporate into 使成为……的一部分
comparable to 可比较的，比得上的
upper right 右上角
call for 要求；需要
be critical for 至关重要，……是关键的

Notes

(1) The main property limitation of metals as structural materials is their low specific strength. Most engineering designs call for structural materials that have high strength, fracture toughness, and stiffness while minimizing weight.

limitation 译为“限制，局限性”；specific strength 译为“比强度”；call for 译为“要求，需要”；that

引导宾语从句，修饰“structural materials”。

参考译文：金属作为结构材料的主要性能限制是其比强度低。大多数工程设计都要求结构材料具有高强度和断裂韧性，在减小重量的同时保持刚度。

(2) Metals can be strengthened through controlled creation of internal defects and boundaries that obstruct dislocation motion. But such strategies compromise ductility and toughness.

controlled 译为“可控的”；dislocation motion 译为“位错运动”；compromise 译为“折中，妥协”。

参考译文：金属可以通过控制内部缺陷的产生和阻碍边界位错运动来强化，但这类策略折中了其延性和韧性。

(3) Another route to enhancing the corrosion resistance of metals is to modify the chemical composition of surface layers. But most such processes require high temperatures that may cause serious deterioration of the metal substrates.

route to 译为“……的方法”；modify 译为“修饰，改性”；that 引导宾语从句。

参考译文：另一种提高金属耐腐蚀性的方法是改变表面层的化学成分。但是大多数这样的工艺需要高温，这可能会导致金属基材的严重退化。

(4) These features are critically important for predicting fracture in engineering structures. In contrast, it remains very difficult to predict fracture of composites and ceramics; their fracture is often catastrophic and may cause serious economic loss and even loss of life.

be critically important for 译为“对……至关重要”；in contrast 译为“相比而言”。

参考译文：这些特征对于预测工程结构的断裂是至关重要的。相比之下，复合材料和陶瓷的断裂预测仍然是非常困难的；它们的断裂往往是灾难性的，并可能导致严重的经济损失，甚至人员伤亡。

(5) Modern technologies not only strongly rely on these unique properties of metals, but urgently call for even better metals. Increasing the strength of metals without sacrificing other properties is critical for their competitiveness.

not only…, but also…译为“不仅……，而且……”；rely on 译为“依赖，依靠”；be critical for 译为“至关重要，……是关键的”。

参考译文：现代技术不仅强烈依赖金属的这些独特特性，而且迫切需要更好的金属。在不牺牲其他性能的情况下增加金属的强度对于它们的竞争力至关重要。

28.2　阅读材料

28.2　Reading Materials

Metal Waste

1. Introduction

The reuse of secondary metals constitutes the oldest and most important kind of recycling activity in the history of humankind. The recycling of metal dates back to approximately 10000 years ago, and the main reason why metals have always been recycled is because of the large amount of energy that must be expended to extract a metal from its ores in the first place. This has not changed over the millennia. Steel and nonferrous metal industries are among those sectors with the highest energy intensities and recycling saves up to 95% of the energy consumption expended in the production of the primary metal. Thus scrap and metal-bearing residues are of great importance when considering the purchase of raw materials to produce virgin metal. This is especially important for those countries which have little or no ore deposits. For instance in Europe, in the field of end-of-life vehicles (ELVs) and waste from electric and electronic equipment (WEEE), the EU directives created additional incentives concerning metal reclamation and recycling.

2. Scrap Metals

Metals for recycling include those that are recovered from large bulky equipment and often consist of both ferrous and nonferrous materials. The metals involved are mainly iron, steel, stainless steel (also called inox-steel), and several nonferrous (NF) metals.

(1) Ferrous Metals

In 2012 the total world steel production amounted to 1.5×10^9 t (1.5 billion tonnes) of which 0.57×10^9 t were fabricated from discarded metal waste. Thus roughly 37% of the world's steel output originates from scrap iron and steel. The energy saved from using recycled steel scrap amounts to approx. 75% of the energy that would have been spent to generate the steel from primary mineral raw materials and the CO_2 emissions are reduced by 58%. Furthermore, the recycling process results in an 86% reduction in air pollutants and a 97% reduction in mining waste.

Stainless steel is an iron alloy containing nickel, chromium (usually it contains a minimum of 10.5% by mass of chromium), and other elements in order to protect the metal against corrosion and other unwanted chemical reactions. The market demand for this metal has doubled over the past decade with an annual production amounting to over 2.5×10^7 t. The recycling of stainless steel is especially important as it saves natural resources including the different alloying metals which are beginning to show signs of future depletion. Today, an ordinary stainless steel product is composed of about 60% recycled material.

(2) Nonferrous Metals

The most commonly used NF metals in the world today are aluminum, copper, brass, zinc, and lead. Because of their limited availability, their high value, and the considerable energy saved if recycled, large quantities of these NF metals are reclaimed and recycled in smelters, refineries, foundries in most countries of the world. This is not only ecologically sound (up to 99% reduction of CO_2 emissions) but also energy efficient (80%~95% savings). The following list of percentages of recycled metals in new products gives some idea of the scale of recycling and its importance: aluminum >33%, copper >40%, zinc >30%, and lead >35%.

3. Management of Metal Waste

The essential demands of modern waste management are to reduce the total amount of waste arising and to reuse and recycle as much of the waste as possible. In many fields of recycling, it is difficult to meet these requirements because the recovered products are often of reduced quality and hence value. As a result, these materials can only be used in a downgraded form.

In most cases the collected material is of mixed types of plastic which are not easy to recycle. Mixed plastics can, however, be used in the production of components with inferior properties. However, for metal recycling the situation is very different. The most important reason for this is that metal can be recycled indefinitely and that subsequent metallurgical treatments do not substantially change the physical and chemical properties of the metal. Furthermore, metals (including recycled metals) have a high market value which is very rare in the recycling industry. Here, with metals, supply and demand are the determining factors that affect the market prices. Moreover, the prices can be influenced to a certain degree by operators of recycling plants as they have the chance to sell their products at convenient points of time, that is, when adequate revenues can be achieved and the "price is right". In general, the recycling industry for metals is structured in an organizational form of a pyramid: at the bottom of the pyramid are many small companies which purchase and collect scrap metal. This scrap is then sold to larger business establishments which process and separate the metals and finally it is sold to companies which have metallurgical plants for further treatment.

4. Metal Containing Raw

Scrap metal is made up of a mixture of metals originating from a variety of sources which include commerce and industry, municipalities, and households. This obsolete scrap is collected, stored, processed, and sold from scrap yards or other specialized facilities. As a rule this type of scrap does not contain foreign material and does not need any treatment or separation process and can be reused directly. However, obsolete scrap requires target-oriented mechanical processing in order to meet the quality demands of the customers which at this stage are metallurgists at metallurgical plants.

The collected scrap appears in a great variety of forms and properties such as different lump sizes and shapes; materials with differing bulk densities; different kinds of metals, each with their own properties (e. g., hardness, abrasiveness, etc.); and materials made of composite substances which could include plastics and other nonmetals. Consequently, it is necessary to adapt the treatment processes to the special characteristics of the most frequently found complex composed feed mixtures. Mechanical processing of scrap is carried out predominantly with mechanical equipment, such as shears, compaction units, shredders, and other types of machines. Table 1 presents an overview of the different kinds of scrap and the mechanical processing methods used in recovering the different metals.

5. Conclusion

Mechanical treatment of different waste streams containing metals involves the use of well-established methods, which as a rule, is profitable because of the high market value of the recovered metals and also because metals do not change their properties with use and hence can be recycled an unlimited number of times. The main aim of the processing methods is to achieve high recovery values as well as the best possible grades of the final metallic products. Sensor based sorters are implemented additionally to improve separation efficiencies. Finally, resmelting of the reclaimed products constitutes the closing of the complete recycling loop for metals.

Table 1 Examples of different scrap types and the possible processing methods employed

Type of Scrap(Fe- and NF-Metals)	Potential Processing Method
Light mixed consumer scrap, end-of-life-vehicles(ELVs)	Comminution with shredder, subsequent separation with air classifier, magnetic separator, and handpicking
Waste of electric and electronic equipment(WEEE)	Comminution with hammer mill, subsequent separation with air classifier, magnetic separator, and other equipment
Cable scrap	Comminution with rotor shears and granulators, subsequent separation with air tables, and other equipment
Ash from waste incineration including Fe- and NF-metals	Classifying with screens, comminution with shredder, subsequent separation of metals
Mixed stainless steel scrap	Comminution with hydraulic shear or shredder, subsequent separation of metals
Intermediate metal products from waste sorting plants	Comminution with hammer mill, subsequent separation with magnetic and eddy current separators, and other equipment
Sheet metals and residues of stamping	Compaction with scrap press
Fe- and NF-turnings	Comminution with turnings crusher, subsequent separation of cutting fluids with centrifuge, magnetic separator
Heavy scrap with wall thicknesses up to 150mm	Comminution with hydraulic scrap shear, where necessary with subsequent screening of fines
Heavy scrap with wall thicknesses more than 150mm	Comminution with flame cutting or blasting
Cast iron scrap	Comminution with vertical drop work

(Source: Trevor M Letcher, et al. *Waste: A Handbook for Management*. Amsterdam: Elsevier, 2011)

Words and Expressions

millennia *n.* 千年
scrap *n.* 碎片，废品，残羹
residue *n.* 剩余物，残留物，余产
depletion *n.* 损耗，耗尽
smelter *n.* 熔炉，冶金厂，熔炼工
foundry *n.* 铸造，铸造类，[机] 铸造厂
shredder *n.* 碎纸机，撕碎者，切菜器
raw material 原材料
virgin metal 原金属，新炼金属

28.3 问题与讨论
28.3 Questions and Discussions

(1) What do you think will replace metals in future commercial jetliner?

(2) What is the main limitation of metals as structural materials?

(3) How to strengthen metals without losing toughness?

(4) How to protect metals from corrosion?

(5) Why metals are irreplaceable?

(6) What is the future development of metals?

参考文献
References

[1] Martin T. Base Metals Handbook[M]. Cambridge:Woodhead Publishing,2006.

[2] Reza A,Lara Abbaschian,Robert E. Physical Metallurgy Principles[M]. Stamford:Cengage Learning,2009.

[3] Senkov O N,Miracle D B,Firstov S A. Metallic Materials with High Structural Efficiency[M]. Holland:Kluwer Academic Publishers,2004.

[4] Ganka Z,Zlatanka M. Imperfections of the Crystal Structure,Microstructure of Metals and Alloys[M]. Boca Raton:CRC Press,2008.

[5] Hull D, Bacon D J. Introduction to Dislocations [M]. Fifth Edition. Burlington: Butterworth-Heinemann, 2011.

[6] Cury S. Mechanical Alloying and Milling[M]. New York:Marcel Dekker,2004.

[7] Adrian P. Mouritz. Introduction to Aerospace Materials[M]. Cambridge:Woodhead Publishing,2012.

[8] Reza A,Lara A,Robert E. Physical Metallurgy Principles[M]. Stamford:Cengage Learning,2011.

[9] Mridha S. Reference Module in Materials Science and Materials Engineering[D]. Glasgow: University of Strathclyde,2016.

[10] Davis J R. Aluminum and Aluminum Alloys-ASM Specialty Handbook[M]. Ohio:ASM International,1993.

[11] Wit G. Advanced Machining Processes of Metallic Materials[M]. Theory Modelling and Applications, Elsevier,2016.

[12] Black J T,Davis J R. American Society for Metals[M]. Ohio:ASM International,1998.

[13] Semiatin S L. Introduction to Forming and Forging Processes[J]. Ohio:ASM Handbook,1993(14).

[14] Siddhartha R. Principles and Applications of Metal Rolling [M]. Cambridge: Cambridge University Press,2016.

[15] Richard E C. Ceramography: Preparation and Analysis of Ceramic Microstructures [M]. Ohio: ASM International,2002.

[16] Banerjee M K. Fundamentals of Heat Treating Metals and Alloys[J]. Comprehensive Materials Finishing, 2017(2):1-49.

[17] Michael J S,Madhu S C. Introduction to Surface Hardening of Steels[J]. ASM Handbook,2013(4).

[18] Robert B H. Plasma-Spray Coating,and Applications[M]. Weinheim: VCH Verlags gesellschaft mbH. 1996.

[19] Adrian P M. Introduction to Aerospace Materials[M]. Cambridge:Woodhead Publishing,2012.

[20] Merhar J. Overview of metal injection moulding[J]. Metal Powder Report,1990,45(5),339-342.

[21] Mittemeijer. Fundamentals of materials science[M]. Berlin:Springer-Verlag Berlin Heidelberg,2010.

[22] Zoltan B,Guido S. Physical Metallurgy[M]. Fifth Edition. Amsterdam: Elsevier publications,2014.

[23] Amit B,Mechanical Properties and Working of Metals and Alloys[M]. Singapore:Springer Nature Singapore Pte Ltd. 2018.

[24] Gene M. The welding of aluminium and its alloys[M]. New York:CRC press LLC & Abington,Woodhead Publishing Limited,2002.

[25] Frederick E Wang. Bonding Theory for Metals and Alloys[M]. Beltsville:Innovative Technology International Inc. ,2005.

[26] Ctirad Uher. Therm Conductivity Theory,Properties,and Applications[M]. Boston:Kluwer Academic/Plenum Publishers,2003.

[27] Philip A,Schweitzer P E,Physical,Mechanical,and Corrosion Properties[M]. Pennsylvania:Marcel Dekker Inc. ,2003.

[28] Miroslav I. Marek[J]. Metals ASM Handbook,1992(13).

[29] Ottewill F G,Barker D. Corrosion and Protection of Metals:Ⅱ. Types of Corrosion and Protection Methods, Article in Transactions of the Institute of Metal Finishing[J]. 1993(71):117-120.

[30] Mahdi Y,Tuan A N,Kenkyu. Journal of Nanotechnology & Nanoscience[J]. 2019(5):37-44.

[31] Beeck O. Catalysis and the Adsorption of Hydrogen on Metal Catalysts[J]. Advances in Catalysis,1950(2): 151-195.

[32] Baddoo N R. Design Manual for Structural Stainless Steel[M]. Berkshire:SCI Publication,2017.

[33] Krishnan K S, Rajiv S M. Metallurgy and Design of Alloys with Hierarchical Microstructures [M]. Amsterdam:Elsevier Inc. ,2017.

[34] Hendra H,Dadan R,Joy R P D. Biomedical Engineering[M]. London:IntechOpen,2011.

[35] Swadhin K P,Biswajit K S. Metallic Glasses:A Revolution in Material Science[M]. London:IntechOpen, 2020.

[36] Clyne T W,Withers P J. An Introduction to Metal Matrix Composites[M]. Cambridge:Cambridge University Press,1995.

[37] Lu K. The Future of Metals[J]. Science,2010,328(5976):319-320.

[38] Alexander F,Thomas P,Jorg J,et al. Waste—A Handbook for Management[M]. London:Academic Press,2019.